Science, Policy, and the Value-Free I

SCIENCE, POLICY, and the VALUE-FREE IDEAL

HEATHER E. DOUGLAS

University of Pittsburgh Press

Q
175
.5
D68

Published by the University of Pittsburgh Press, Pittsburgh, Pa., 15260
Copyright © 2009, University of Pittsburgh Press
All rights reserved
Manufactured in the United States of America
Printed on acid-free paper
10 9 8 7 6 5 4 3 2 1

Library of Congress Cataloging-in-Publication Data

Douglas, Heather E.
　Science, policy, and the value-free ideal / Heather E. Douglas.
　　　p.　　cm.
　Includes bibliographical references and index.
　ISBN-13: 978-0-8229-6026-3 (pbk. : alk. paper)
　ISBN-10: 0-8229-6026-5 (pbk. : alk. paper)
　1. Science—Social aspects. 2. Science—Moral and ethical aspects. 3. Scientists—Professional ethics. I. Title.
　　Q175.5.D68 2009
　　174'.95—dc22　　　　　　　　　　　　　　　　2009005463

CONTENTS

LIST OF ABBREVIATIONS   VII

PREFACE   IX

CHAPTER 1. Introduction: Science Wars and Policy Wars   1

CHAPTER 2. The Rise of the Science Advisor   23

CHAPTER 3. Origins of the Value-Free Ideal for Science   44

CHAPTER 4. The Moral Responsibilities of Scientists   66

CHAPTER 5. The Structure of Values in Science   87

CHAPTER 6. Objectivity in Science   115

CHAPTER 7. The Integrity of Science in the Policy Process   133

CHAPTER 8. Values and Practices   156

EPILOGUE   175

NOTES   179

REFERENCES   193

INDEX   205

ABBREVIATIONS

AAAS	American Association for the Advancement of Science
AEC	Atomic Energy Commission
CPSC	Consumer Product Safety Commission
DOD	Department of Defense
EPA	Environmental Protection Agency
FACA	Federal Advisory Committee Act
FCST	Federal Council for Science and Technology
FDA	Food and Drug Administration
GAC	General Advisory Committee
IPCC	Intergovernmental Panel on Climate Change
NACA	National Advisory Committee on Aeronautics
NDRC	National Defense Research Committee
NIH	National Institutes of Health
NRC	National Research Council
NRB	National Resources Board
NCB	Naval Consulting Board
NSF	National Science Foundation
OSHA	Occupational Safety and Health Administration
ODM	Office of Defense Mobilization
OMB	Office of Management and Budget
ONR	Office of Naval Research
OST	Office of Science and Technology
OSRD	Office of Scientific Research and Development
PSAC	Presidential Science Advisory Committee
SAB	Science Advisory Board
SAC	Science Advisory Committee

PREFACE

This book has been a long time in the making. I first conceived of the project in 2001 as an arc through the historical, philosophical, and practical terrain of science in policymaking. It seemed to me that the chronic debates and misconceptions that plague this terrain stemmed from the embrace of a particular ideal for scientific reasoning, the value-free ideal. Articulated clearly and defended by philosophers of science for over forty years, it was also pervasive in the science policy communities with which I was in conversation, particularly the risk assessment community I found at the Society for Risk Analysis (SRA). At the SRA's annual meetings, I found not only a dynamic and open set of people deeply committed to hashing out the scientific implications of toxicology, epidemiology, biochemistry, and other disciplines, but also a community that believed that social and ethical values were not supposed to be involved in the assessment of this science, even though they continually found themselves unable to do a complete assessment without those values. The tensions were palpable, even as the fundamental norms causing those tensions were not made explicit. I wanted to bring those norms out in the open, examine their historical roots, see if they were in fact the correct norms, and, if possible, attempt to resolve the tensions.

This book is the result. As a consequence, I began the project with three distinct audiences in mind. First, the book was to make a contribution to the philosophy of science, for that is the discipline that articulates most clearly and guards most zealously the norms of science. If I could not make arguments that were at least provocative, if not convincing, to this community, I would doubt the reliability of my own arguments. However, as I delved into the material, I began to see more clearly the historical roots of the discipline of philosophy of science, and how the value-free ideal was foundational to its very self-conception. In many ways, the arguments here will challenge philosophers of science over what philosophy of science *is* or should be, as the arguments suggest that an approach to the topic that is focused on the purely epistemic will always be inadequate. Science is more than an epistemic enterprise; it is also a moral enterprise and philosophers of science ignore this at their peril.

The book is also written for scientists. I hope I have kept the philosophical jargon and typically turgid writing to a minimum, so that scientists enjoy the read, and find it helpful in thinking about the tensions they face every day in their practices. It is not an easy time to be a scientist. The relationship between science and society has become increasingly fraught with tension, the sources of science funding have shifted dramatically, and it is not always the case that critiques of science arise from ignorance of science and thus can be attributed to scientific illiteracy. What can and should be expected of scientists is not as clear as it once was (although I suspect that this issue has always been somewhat contested). I hope that this book helps to clarify some key expectations for scientists, as well as provide some useful guidance in an increasingly complex world.

Finally, the book is written for policymakers and for anyone interested in policymaking. For too long, the policy process has been hamstrung by inappropriate expectations of what science can provide. Commentators have bemoaned the expectation that science could provide definitive guidance for policymakers, and they have complained that policymakers have sought the science that would support their predetermined policy choices, ignoring other evidence. Both extremes are abuses of science, ways of utilizing the prima facie authority of science to cover up the need for difficult choices in policymaking. But as this book should make clear, science is not the value-neutral terrain that policymakers might desire, and any use of science must acknowledge the value choices embedded in that use, even accepting a scientific claim as adequately supported by a body of evidence. Hopefully, understanding this will short-circuit much of the fruitless sound science–junk science rhetoric from the past two decades, making the way for a more transparent and open use of science for policymaking.

Whether this book can successfully speak to all these audiences remains to be seen. Perhaps it was too much to expect of one book, but I am loath to write the same book more than once. Instead, I took a long time in writing it once. Many people have provided invaluable support throughout the process. I began the project in earnest with a year's sabbatical funded both by the University of Puget Sound's Martin Nelson Junior Sabbatical Fellowship and by a National Science Foundation Societal Dimensions of Engineering, Science, and Technology Grant, #0115258. Without that year, I doubt this project would have ever gotten off the ground, and my colleagues at the University of Puget Sound provided useful sounding boards for some of the early forays into the key philosophical ideas.

Audiences at conferences also helped to refine the main arguments, particularly conferences such as the Pittsburgh-Konstanz 2002 Colloquium in the Philosophy of Science on Science, Values, and Objectivity at the University of Pittsburgh (many thanks to Sandra Mitchell for a commentary that spurred deeper reflection), the 2003 Yearbook of the Sociology of the Sciences Conference, "On Scientific Expertise and Political Decision-Making," at Basel, Switzerland (which helped me hone how to think and write about the public's role in risk analysis), the Chemical Heritage Foundation's 2006 workshop, "Towards a History and Philosophy of Expertise," the 2006 workshop on "Evidence and Dissent in Science" at the London School of Economics, and at meetings of the Philosophy of Science Association, the American Philosophical Association, the International Society for the History of Philosophy of Science, the American Association for the Advancement of Science, the Society for the Social Studies of Science, and, of course, the Society for Risk Analysis. In addition, I have been fortunate enough to have given talks at university colloquia that have afforded me captive and critical audiences at the University of British Columbia, the University of Tennessee, the University of North Texas, University of California at Santa Cruz, the University of Utah, the University of Bielefeld, and Arizona State University. At all these venues, colleagues and critics too numerous to mention have provided essential feedback.

Special thanks to the History and Philosophy of Science and Technology Reading Group (aka Hipsters) at the University of Tennessee for detailed comments on chapters 1 and 5, to Gary Hardcastle for comments on chapters 2 and 3, to Hasok Chang for comments on chapter 7, to Nancy Cartwright for her interest in and support of the whole book, to David Guston for encouragement throughout, and to the folks at the Oak Ridge Institute for Continued Learning for in-depth discussions of chapters 3–6. George Reisch, John Beatty, Alan Richardson, and Don Howard have provided crucial insights on the history of philosophy of science, a key part of this account. Discussions with Janet Kourany in 2002 led to many of the insights of chapter 5; fruitful debates with Hugh Lacey have also furthered those arguments. Elijah Millgram has been an invaluable conversant on the core philosophical ideas. My colleagues from the Society for Risk Analysis, including Katy Walker, Jim Wilson, Lorenz Rhomberg, Resha Putzrath, Steve Lewis, and Rick Belzer, have provided a continual dose of reality about the science in policy process, even if that is not reflected fully in this book. Also, thanks go to my colleagues at the University of Tennessee for helping me to negotiate the publishing process. The comments from all the blind reviewers of this manuscript have certainly helped make this a better book. Thanks also to my copy editor, Kathy McLaughlin, University of Pittsburgh Press director Cynthia Miller, and all the staff at the press for helping me to bring this project to fruition. Finally, Peter Machamer deserves a warm thanks for being willing to take a chance on a rather crazy idea I had as a graduate student, to write a dissertation on the role of science in policymaking. I would have become a rather different philosopher of science, if I pursued my Ph.D. at all, without that start.

Most importantly, this book would not have been possible without the love and support of my partner in philosophy and all other things, Ted Richards. He has read the manuscript in various forms more times than I can count, helped me clarify the arguments, seen crucial flaws, and even helped me format it. He has both put me on planes and made home a wonderful place to which I can return. And he has encouraged me when I needed it most. His patience and perseverance have been essential and unfathomable, and to him this book is dedicated.

Science, Policy, and the Value-Free Ideal

CHAPTER 1

INTRODUCTION
Science Wars and Policy Wars

WHEN CONSIDERING THE IMPORTANCE of science in policymaking, common wisdom contends that keeping science as far as possible from social and political concerns would be the best way to ensure science's reliability. This intuition is captured in the value-free ideal for science—that social, ethical, and political values should have no influence over the reasoning of scientists, and that scientists should proceed in their work with as little concern as possible for such values. Contrary to this intuition, I will argue in this book that the value-free ideal must be rejected precisely because of the importance of science in policymaking. In place of the value-free ideal, I articulate a new ideal for science, one that accepts a pervasive role for social and ethical values in scientific reasoning, but one that still protects the integrity of science.

Central to the concerns over the use of science in policymaking is the degree of reliability we can expect for scientific claims. In general, we have no better way of producing knowledge about the natural world than doing science. The basic idea of science—to generate hypotheses about the world and to gather evidence from the world to test those hypotheses—has been unparalleled in producing complex and robust knowledge, knowledge that can often reliably guide decisions. From an understanding of inertia and gravity that allows one to predict tides and the paths of cannonballs, to an understanding of quantum mechanics that underlies the solid state components of computers, to an understanding of physiology that helps to guide new medical breakthroughs, science has been remarkably successful in developing theories that make reliable predictions.

Yet this does not mean that science provides certainty. The process of hypothesis testing is inductive, which means there is always a gap between the evidence and the theory developed from the hypothesis. When a scientist makes a hypothesis, she is making a conjecture of which she is not certain. When the gathered evidence supports the hypothesis, she is still not certain. The evidence may support the theory or hypothesis under examination, but there still may be some other theory that is also supported by the available evidence, and more evidence is needed to differentiate between the two. The hypothesis concerns a great many more instances than those for which we will carefully collect data. When we collect more data, we may find that seemingly well-confirmed hypotheses and theories were false. For example, in the late nineteenth century, it was widely accepted that chemical elements could not transform into other elements. Elements seemed to be stable in the face of any efforts at transmutation. The discovery of radioactivity in the early twentieth century overturned this widespread belief. Or consider the theory of ether, a medium in which it was once commonly believed light traveled. Despite near universal acceptance in the late nineteenth century, the theory of ether was rejected by most physicists by 1920. Going even further back in history, for over 1,500 years it seemed a well-supported theory that the sun revolved around the Earth, as did the fixed stars. But evidence arose in the early seventeenth century to suggest otherwise and, along with changes in the theories of mechanics, overturned one of the longest standing and best supported scientific theories of the time. After all, how many times had humans seen the sun rise and set? And yet, the theory was ultimately incorrect. Data can provide evidential support for a theory, but can never prove a scientific theory with certainty. Aspects of the world that were once thought to be essential parts of scientific theory can be rejected wholesale with the development of new theories or the gathering of new evidence.

Because of the chronic, albeit often small, uncertainty in scientific work, there is always the chance that a specific scientific claim is wrong. And we may come to know that it is wrong, overturning the theory and the predictions that follow from it. The constant threat of revision is also the promise of science, that new evidence can overturn previous thought, that scientific ideas respond to and change in light of new evidence. We could perhaps have certainty about events that have already been observed (although this too could be disputed—our descriptions could prove inaccurate), but a science that is only about already observed events is of no predictive value. The generality that opens scientific claims to future refutation is the source of uncertainty in science, and the source of its utility. Without this generality, we could not use scientific theories to make predictions about what will happen in the next case we encounter. If we want useful knowledge that includes predictions, we have to accept the latent uncertainty endemic in that knowledge.

The chronic incompleteness of evidential support for scientific theory is no threat to the *general* reliability of science. Although we can claim no certainty for science, and thus no perfect reliability, science has been stunningly successful as the most reliable source for knowledge about the world. Indeed, the willingness to revise theories in light of new evidence, the very quality that makes science changeable, is one key source for the reliability and thus the authority of science. That it is not dogmatic in its understanding of the natural world, that it recognizes the inherent incompleteness of empirical evidence and is willing to change when new evidence arises, is one of the reasons we should grant science a prima facie authority.

It is this authority and reliability that makes science so important for policy. And it seems at first that the best way to preserve the reliability of science is to keep it as far from policy as possible. Indeed, the realm of science and the realm of policy seem incompatible. In the ideal image of science, scientists work in a world detached from our daily political squabbles, seeking enduring empirical knowledge. Scientists are interested in timeless truths about the natural world rather than current affairs. Policy, on the other hand, is that messy realm of conflicting interests, where our temporal (and often temporary) laws are implemented, and where we craft the necessary compromises between political ideals and practical limits. This is no place for discovering truth.

Without reliable knowledge about the natural world, however, we would be unable to achieve the agreed upon goals of a public policy decision. We may all agree that we want to reduce the health effects of air pollution, for example, or that we want safe, drinkable water, but without reliable information about which pollutants are a danger to human health, any policy decision would be stymied in its effectiveness. Any implementation of our policy would fail to achieve its stated goals. Science is essential to policymaking if we want our policies concerning the natural world to work.

This importance of science in achieving policy goals has increased steadily throughout the past century in the United States, both as the issues encompassed by public policy have expanded and as the decisions to be made require an increasingly technical base. As science has become more important for policy, the relationship between science and policy has become more entangled. This entanglement exists in both directions: science for policy and policy for science. In the arena of policy for science, public funds allocated for doing science have grown dramatically, and these funds require some policy decisions for which projects get funded and how those funds will be administered. In the arena of science for policy, increasing numbers of laws require technically accurate bases for the promulgation of regulations to implement those laws. These arenas in practice overlap: which studies one chooses to pursue influences the evidence one has on hand with which to make decisions. In this book, however, my focus will be largely on science for policy.

While the entanglement between science and policy has been noted, the importance of this entanglement for the norms of science has not been recognized. As science plays a more authoritative role in public decision-making, its responsibility for the implications of research, particularly the implications of potential inductive error, increases. Failure to recognize the implications of this responsibility for science, combined with the desire to keep science and policy as distinct as possible, has generated deep tensions for our understanding of science in society.

These tensions are evident in the increased stress science has been under, particularly with respect to its public role. Some commentators note an increasing strain on the "social contract" between science and society (see, for example, Guston and Keniston 1994). This strain was made manifest in the 1990s when two public debates erupted over science: the "Science Wars" and the sound science–junk science dispute. Both can be taken as emblematic of science under stress in our society.

The Science Wars, as they are often called, centered on the authority of science. They were about whether or not science should be believed when it tells us what the nature of the world is, about whether or not science should have more public authority than other approaches to knowledge or belief. For those outside the world of science studies, these are astonishing questions to raise. If one wants to know something about the natural world, it seems obvious that one should ask scientists. While few in science studies would actually dispute this, the claim has been made that the knowledge produced by science has no special authority above and beyond any other approach. In other words, the claim is that science and its methods have no special hold on the ability to uncover and speak truth; they simply have more funding and attention.

The sound science–junk science war, in contrast, does not question the special epistemic authority given to science in general, or the overall reliability of science for answering empirical questions. Instead, this dispute is about which particular piece(s) of science should shape policy. When is a particular body of scientific work adequately "sound" to serve as the basis for policy? Debates in this arena center on how much evidence is sufficient or when a particular study is sufficiently reliable. The arguments focus on such questions as: How much of an understanding of biochemical mechanisms do we need to have before we regulate a chemical? How much evidence of causation is needed before a court case should be won? How much of an understanding of complex biological or geological systems do we need before regulatory frameworks intervene in the market to prevent potential harm? The idea that science is the authoritative body to which one should turn is not questioned; what is questioned is which science is adequate for the job, or which scientific experts are to be believed by policymakers, Congress, and the public.

While both of these disputes are symptomatic of deep concerns surrounding the public role of science, neither has been able to produce a satisfactory approach to understanding the role of science in society or what that role might mean for the norms of scientific reasoning. This is, in part, because both disputes began with the presupposition that science is a distinct and autonomous enterprise developed by a community of scientists largely in isolation from public questions and concerns. Such an understanding of science and scientists inhibits a clear view of how science should function in society. Both in the academic arena of the Science Wars and in the policy arena of the sound science–junk science dispute, the discussions shed little light on the deep questions at issue, even as the existence of the debates indicated the need for a more careful examination of the role of science in society and its implications.

The Science Wars

The Science Wars were an academic affair from start to finish. A particular critique of science, known as social constructivism, began in the 1970s and gathered steam and fellow travelers in the 1980s. The social constructivist critique was essentially an assault on the authority of science, particularly its apparently privileged place in producing knowledge. Social constructivists suggested that scientific *knowledge* (not just scientific institutions or practices) was socially constructed and thus should be treated on a par with other knowledge claims, from folklore to mythology to communal beliefs

(Barnes and Bloor 1982). There simply was no deep difference between one set of knowledge claims and another, social constructivists argued, and thus scientific facts held no special claim to our acceptance.

As this critique was developed throughout the late 1970s and 1980s, other criticisms of science began to coalesce. For example, feminists noted that few scientists were women, and that many scientific claims about women had been (and continued to be in the 1980s) either explicitly sexist or supportive of sexist beliefs (Fausto-Sterling 1985; Longino 1990). Feminists wondered if science done by women would be different, producing different conclusions (Harding 1986, 1991). It was unclear whether sexist science was always methodologically flawed or bad science (as it sometimes was), or whether sexist science simply relied upon different background assumptions, assumptions which in themselves did not clearly put the scientific quality of the work in doubt. If the latter were the case, then an emphasis on unpacking the background assumptions, which often arose from the surrounding culture, seemed to support the notion that science was in fact a social construct, or at least heavily influenced by the surrounding society and its prejudices. Although feminists and social constructivists disagreed about much, their arguments often pointed in a similar direction—that scientific knowledge consisted of socially constructed claims that were relative to a social context. Only those *within* a particular social context thought the claims produced had any special authority or believability.

By the early 1990s, some scientists began to take umbrage with these criticisms, particularly the apparently strong form of the social constructivist critique, that in general science had no special claim to being more believable than any other knowledge claim. As scientists began to engage in this debate, the Science Wars erupted. An early salvo was Lewis Wolpert's *The Unnatural Nature of Science* (1992), which devoted a chapter to responding to relativist and social constructivist views of science. The debate really heated up in 1994, however, with the publication of Paul Gross and Norman Levitt's *Higher Superstition: The Academic Left and Its Quarrels with Science*.[1] As one sympathetic reader of the book notes, "This unabashedly pugnacious work pulled no punches in taking on the academic science critics. . . . Naturally, those criticized on the 'academic left' fired back, and so the science wars were joined" (Parsons 2003, 14). The polemical nature of Gross and Levitt's book drew immediate attention from scientists and fire from its targets, and the accuracy of Gross and Levitt's criticisms has been seriously questioned. (Roger Hart [1996] is particularly precise in his critique of Gross and Levitt for simply misunderstanding or misrepresenting their targets.) Now scientists and their critics had a text over which to argue.

The Science Wars took an even nastier turn when Alan Sokal, a physicist, decided to attempt a hoax. Inspired by Gross and Levitt's book, he wrote a paper in the style of postmodern social constructivism and submitted it for publication in a left-leaning social constructivist journal, *Social Text*. The paper was entitled "Transgressing the Boundaries: Toward a Transformative Hermeneutics of Quantum Gravity," and was a parody of some constructivist work, citing and drawing heavily from that work. The editors were thrilled that a physicist was attempting to join in the discussion, and they published the piece in 1996.[2] Sokal then revealed he had written the work as a hoax to unmask the vacuity of this kind of work (see Sokal 1998). Many cheered Sokal's effort; after all, hoaxing is a venerable tradition in the natural sciences, where hoaxing has revealed some of science's most self-deceived practitioners.[3] But in the humanities, there is little tradition of hoaxing as a deliberate attempt to catch a colleague's suspected incompetence.[4] Scholars in those fields take for granted that a person's work, ingenuously put forth, is their own honest view, so others cried foul at Sokal's violation of this basic norm of intellectual honesty. The gulf between the critics of science and the scientists only grew wider.

However, as Ullica Segerstråle notes, in many of the forums of debate for the Science Wars, it was hard to find anyone actually defending the strong versions of the social constructivist claims (Segerstråle 2000, 8). The plurality of views about science among both science studies practitioners and scientists themselves became increasingly apparent as the decade came to a close. In the end, the Science Wars petered out, perhaps having moderated the views of some academics, but having had little impact on the public perception or understanding of science.[5]

So what was the debate about? Why did critics of science attack science's general authority? I think the debate arose because of tensions between science's authority and science's autonomy. As I will discuss in chapter 3, the autonomy of science, the isolation of science from society, became a cornerstone of the value-free ideal in the 1960s. On the basis of the value-free nature of science, one could argue for the general authoritativeness of its claims. But an autonomous *and* authoritative science is intolerable. For if the values that drive inquiry, either in the selection and framing of research or in the setting of burdens of proof, are inimical to the society in which the science exists, the surrounding society is forced to accept the science and its claims, with no recourse. A fully autonomous and authoritative science is too powerful, with no attendant responsibility, or so I shall argue. Critics of science attacked the most obvious aspect of this issue first: science's authority. Yet science is stunningly successful at producing accounts of the world. Critiques of science's general authority in the face of its obvious importance seem absurd. The issue that requires serious examination and reevaluation is not the authority of science, but its autonomy. Simply assuming that science should be autonomous, because that is the supposed source of authority, generates many of the difficulties in understanding the relationship between science and society.

That the relationship between science and society was an underlying but unaddressed tension driving the Science Wars has been noted by others. Segerstråle, in her reflections on the debate, writes, "But just as in other academic debates, the issues that were debated in the Science Wars were not necessarily the ones that were most important. One 'hidden issue' was the relationship between science and society. The Science Wars at least in part reflected the lack of clarity in science's basic social contract at the end of the twentieth century" (Segerstråle 2000, 24–25). Rethinking that social contract requires reconsidering the autonomy of science. Once we begin to rethink the autonomy of science (chapter 4), we will need to rethink the role of values in science (chapter 5), the nature of scientific objectivity (chapter 6), and the process of using science in policymaking (chapter 7). The Science Wars demonstrated the tension around these issues with its intensity, but shed little light on them.

Policy Wars: The Sound Science–Junk Science Dispute

While the Science Wars were playing out, the place of science in policymaking was the focus of a completely separate debate in the 1990s. Rather than centering on the authority of science in general, the debate over science in policy centered on the reliability of particular pieces of science. As noted above, science is endemically uncertain. Given this uncertainty, blanket statements about the general reliability of science, and its willingness to be open to continual revision, are no comfort to the policymaker. The policymaker does not want to be reassured about science in general, but about a particular piece of science, about a particular set of predictions on which decisions will be based. Is the piece of science in which the policymaker is interested reliable? Or more to the point, is it reliable enough? For the policymaker to not rely on science is unthinkable. But which science and which scientists to rely upon when making policy decisions is much less clear.

The difficulties with some areas of science relevant for policy are compounded by their complexity and by problems with doing definitive studies. Because of their complexity, policy interventions based on scientific predictions rarely provide good tests of the reliability of the science. A single change in policy may not produce a detectable difference against the backdrop of all the other factors that are continually changing in the real world. And the obvious studies that would reduce uncertainty are often immoral or impractical to perform. For example, suppose we wanted to definitively determine whether a commonly occurring water pollutant caused cancer in humans. Animal studies leave much doubt about whether humans are sufficiently similar to the animal models. The biochemical mechanisms are often too complex to be fully traced, and whether some innate mechanism exists to repair potential damage caused by the pollutant would be in doubt. Epidemiological studies involve people living their daily lives, and thus there are always confounding factors, making causation difficult to attribute. A definitive study would require the sequestering of a large number of people and exposing them to carefully controlled doses. Because the latency period for cancer is usually a decade or more, the subjects would have to remain sequestered in a controlled environment for years to avoid confounders. And large numbers of people would be needed to get statistically worthwhile results. Such a study would be immoral (the subjects would not likely be volunteers), expensive, and too unwieldy to actually conduct. We cannot reduce uncertainty by pursuing such methods. Nor does implementing a policy and seeing what happens reduce uncertainty about the science. Real world actions are subject to even more confounders than are controlled for in epidemiological studies. Even if cancer rates clearly dropped after the regulation of a pollutant, it would be hard to say that the new regulation caused the drop. Other simultaneous regulations or cultural changes could have caused the cancer rate to decline at the same time as the drop in exposure to the pollutant.

Thus, in addition to the generic uncertainties of science, science useful for policymaking often carries with it additional sources of uncertainty arising from the biological, ecological, and social complexity of the topics under study. Yet this chronic uncertainty does not mean that policymakers should go elsewhere for information. Instead, it puts increased pressure on assuring that the science on which they depend is reliable.

Tensions over the role of science in policy increased as the reach of regulation grew and the decisions stakes rose throughout the 1970s and 1980s. As will be discussed in the next chapter, by the 1970s dueling experts became a standard phenomenon in public policy debates. As I will discuss in chapter 7, attempts to proceduralize the policy decisionmaking process were supposed to rein in the impact of these contrarian experts, but by the 1990s it was apparent the problem was not going away. In fact, it seemed to be worsening, with public debates about technical matters occurring increasingly earlier in the policymaking process.

New terms arose in attempts to grapple with the crisis. Although scientists had used the phrase "sound science" to refer to well-conducted, careful scientific work throughout the twentieth century, its opposite was often "pseudoscience," which masqueraded as science but had not a shred of scientific credibility. Examples of pseudoscience included tales of extraterrestrial sightings, claims about extrasensory perception or psychic abilities, and astrology. Pseudoscience as such has not been taken seriously in the policy realm and has not been a source of policy controversy. Rather than these more outlandish concerns, the realm of policy was focused on the reliability of a range of ostensibly reasonable scientific claims. Even as scientific expertise became the basis of many decisions, from new regulatory policy to rulings in tort cases, increasing concern was raised over the quality of science that served as a basis for those decisions. As tension brewed over the role of science in policymaking and skeptics over certain uses of science became more vocal in the early 1990s, a new term entered into the lexicon of science policy commentators: "junk science."

Although the term "junk science" appeared occasionally before 1991, Peter Huber's *Galileo's Revenge: Junk Science in the Courtroom* popularized the notion of junk science. Huber's critique centered on the use of evidence in tort cases and decried the shift in the Federal Rules of Evidence from the Frye rule of the 1920s, which stated that only scientific ideas reflecting the consensus of the scientific community were admissible in court, to more recent, laxer standards that allow any expert testimony that assists in the understanding of evidence or determination of fact. Huber argued that this was a catastrophe in the making and that we needed to strengthen standards back to the Frye rule.

Huber relied upon an autonomous image of science to decide what counted as sound science, stating near the close of his book, "as Thomas Kuhn points out, a scientific 'fact' is the collective judgment of a specialized community" (Huber 1991, 226). For Huber, what the specialized community deems sound science *is* sound science. Yet Kuhn's idea of a specialized community consists of scientists working within internally determined paradigms—sets of problems and ideas that scientists alone, separated from any social considerations, decide are acceptable. (Kuhn's influence in this regard will be discussed further in chapter 3.) Under this Kuhnian image of science as isolated and autonomous, one could presume that there might exist a clear and "pure" scientific consensus to which one could refer, and on which one could rely. Any science outside of this clear consensus was "junk," even if later it might prove its mettle. Initially, the very idea of junk science depended on an autonomous and isolated scientific community, inside of which one could find sound science, and outside of which lay junk science.

Thus, the same conceptual framework that led to the Science Wars, the idea of science as autonomous and isolated, shaped the sound science–junk science debates. Like the Science Wars, the resulting debates in the policy arena have produced few helpful insights. Instead, they merely changed the rhetoric of policy disputes. As experts with obvious credentials continued to disagree about apparently scientific matters, Huber's term "junk science" expanded from the courtroom to all public debates over technical policy issues. Rather than argue that an opposing view had an insufficient scientific basis, one could dismiss an opponent by claiming that their views were based on junk science, which looked like science, but which would be proven wrong in the near future. Conversely, one's own science was sound, and thus would prove to be a reliable basis for decisionmaking in the long run.

The idea that sound science was a clear and readily identifiable category, and that its opposite, junk science, was also easily identified, ran rampant through public discussions. As this language permeated policy debate, it became a mere rhetorical tool to cast doubt upon the expertise of one's opponents. In a revealing study by Charles Herrick and Dale Jamieson, the use of the term "junk science" in the popular media from 1995 to 2000 was examined systematically (Herrick and Jamieson 2001). They found that the vast majority of studies tarnished with the term did not have any obvious flaws (such as lack of peer review or appropriate publication, lack of appropriate credentials of the scientists, or fraud), but were considered "junk science" because the implications of the studies were not desirable. For example, studies were called junk science because the results they produced were not "appropriately weighted" when considered with other evidence, the studies came from a source that was simply presumed to be biased, or undesirable consequences that might follow from the study were not considered. Thus, by the end of the decade, the term "junk science" had come to be used in ways quite different from the original intent of designating work that fails to pass muster inside the scientific community, denoting instead science that one did not like rather than science that was truly flawed.

Despite the muddling of the notions of sound and junk science, much effort has gone into finding ways to sort the two out in the policy process. For example, the Data Quality Act (or Information Quality Act, Public Law 106-554, HR 5658, sec. 515) was passed in 2000 and charged the Office of Management and Budget (OMB) with ensuring "the quality, objectivity, utility, and integrity of information . . . disseminated by Federal agencies," including the information that serves as a basis in public record for regulatory decisions.[6] However, there are deep tensions generally unrecognized at the heart of such solutions to the sound science–junk science problem. In an essay by Supreme Court Justice Stephen Breyer, published in *Science*, Breyer emphasized the need for "sound science" in many then-current legal cases: "I believe there is an increasingly important need for law to reflect sound science" (Breyer 1998, 538). While the importance of sound science is clear, how to identify what constitutes sound science in any particular case is a challenge, Breyer acknowledged. This is in part because the ideal for sound science contains contradictory impulses, as can be seen in Breyer's concern that "the law must seek decisions that fall within the boundaries of scientifically sound knowledge and approximately reflect the scientific state of the art" (Breyer 1998, 537). As noted above, earlier standards of evidence, following the Frye rule, demanded that scientific testimony reflect the consensus of the scientific community. While such a standard might clearly determine the boundaries of scientifically sound knowledge, it would often exclude state-of-the-art science, which would encompass newer discoveries still in the process of being tested and disputed by scientists. Every important discovery, from Newton's theory of gravity to Darwin's descent by natural selection to Rutherford's discovery of radioactivity, was disputed by fellow scientists when first presented. (Some may note that in high stakes discoveries, expert disputation can become a career unto itself.) Yet many cutting-edge scientists have strong evidence to support their novel claims. State-of-the-art science and scientific consensus may overlap, but they are not equivalent. If we want to consider state-of-the-art scientific work in our decisionmaking, we will likely have to consider science not yet part of a stalwart consensus.

In the 2000s, the rhetoric around science in policy changed again, this time to focus on "politicized science" rather than junk science. The Bush administration's handling of science and policy led to these charges, particularly as concern over the suppression of unwanted scientific findings arose.[7] Rather than introducing junk science into the record, the worry is that sound science is being distorted or kept out of the public record altogether. Thus, the debate over the role of science in policymaking continues, even if under an altered guise. Regardless of the form it takes, debate over sound science and junk science (or politicized science) centers on the reliability of science to be used in decisionmaking.

The introduction of new jargon, however, has not helped to clarify the issues. As with the Science Wars, more heat than light has resulted. And ironically, despite the parallels between the two disputes, neither dispute seems to have noticed the other. The Science Wars were a debate among academics interested in science and science studies; the sound science–junk science dispute is a debate among those interested in the role of science in policy and law. One was about the standing of science in society; the other is about which science should have standing. These two disputes involve different texts, participants, and issues, and we should not be surprised that no general connection was made between them. Yet the origins of these two disputes can be found in the same set of historical developments, the same general understanding of science and its place in society. Both disputes and their conceptual difficulties arise from assuming that a clearly defined, authoritative, and autonomous scientific community that hands to society fully vetted scientific knowledge is the correct understanding of science's role in society. Getting to the heart of this understanding—centering on the autonomous and authoritative view of science—will be central to finding a workable resolution to the continuing dispute over the role of science in public policy. It will also challenge the norms for scientific reasoning in general.

Overview, Context, and Limits of the Book

This book will not challenge the idea that science is our most authoritative source of knowledge about the natural world. It will, however, challenge the autonomy of science. I will argue that we have good grounds to challenge this autonomy, particularly on the basis of both the endemic uncertainty in science and science's importance for public decisionmaking. In order to protect the authority of science without complete autonomy, I will articulate and defend ways to protect the integrity of science even as scientific endeavors become more integrated with the surrounding society. By considering carefully the importance of science for public policy, I will argue for important changes in the norms that guide scientific reasoning. In particular, I will argue that the value-free ideal for science, articulated by philosophers in the late 1950s and cemented in the 1960s, should be rejected, not just because it is a difficult ideal to attain, but because it is an undesirable ideal.[8] In its place, I will suggest a different ideal for scientific integrity, one that will illuminate the difference between sound science and junk science, and clarify the importance and role for values in science. I will also argue that rejecting the value-free ideal is no threat to scientific objectivity. With these conceptual tools in hand, a revised understanding of science in public policy becomes possible. I will argue that understanding scientific integrity and objectivity in the manner I propose allows us to rethink the role of science in the policy process in productive ways, ways that allow us to see how to better democratize the expertise on which we rely, without threatening its integrity.

Key to this account is the growth in science advising in the United States. Prior to World War II, involvement of science with government was sporadic. Wartime, such as World War I, produced spurts of activity, but rather than producing a long lasting science-government relationship, these episodes developed the forms of the relationship that would be cemented after World War II. That war was the watershed, when science established a permanent relationship with government, both as a recipient of federal support and as a source for advice. Yet the road since World War II has not been smooth. Chapter 2 will detail both how the forms of science advice originated and the ups and downs of science advising since then. Although the specific avenues for advising have shifted in the past fifty years, the steadily expanding importance of science for policymaking will be apparent.

This continual expansion is crucial to note because even as scientists were becoming more central figures in policymaking, philosophers of science were formulating an understanding of science that would turn a blind eye toward this importance. Chapter 3 will examine how the idea of the science advisor came to be excluded from the realm of philosophy of science. In particular, I will examine how the current ideal for value-free science came into existence. Although some normative impulse to be value-free has been part of the scientific world since at least the nineteenth century, the exact form of the value-free ideal has shifted. At the start of World War II, most prominent philosophers rejected the older forms of the ideal as unworkable. The pressures of the cold war and the need to professionalize the young discipline of philosophy of science generated a push for a new value-free ideal, one that was accepted widely by the mid-1960s, is still predominant among philosophers, and is reflected by scientists. I will describe how this ideal came into existence and how it depends crucially on a belief in the autonomy of science from society.

Chapter 4 begins the critique of this value-free ideal. As we will see in chapter 3, the current value-free ideal rests on the idea that scientists should act as though morally autonomous from society, in particular that they should not consider the broader consequences of their work. Chapter 4 disputes this claim, arguing that scientists must consider certain kinds of consequences of their work as part of a basic responsibility we all share. Because of this responsibility, the value-free ideal cannot be maintained. Values, I argue, are an essential part of scientific reasoning, including social and ethical values.

This raises the question of how values should play a role in science, a question addressed in chapter 5. There I lay out a normative structure for how values should (and should not) function in science, and I argue that at the heart of science values must be constrained in the roles they play. The crucial normative distinction is not in the kinds of values in science but in how the values function in the reasoning process. While no part of science can be held to be value-free, constraints on how the values are used in scientific reasoning are crucial to preserving the integrity and reliability of science. By clearly articulating these constraints, we can see the difference between acceptable science and politicized science, between sound and junk science.

If science is and should be value-laden, then we need an account of objectivity that will encompass this norm, an account that explicates why we should trust specific scientific claims and what the bases of trust should be. In chapter 6 I provide that account, arguing that there are at least seven facets to objectivity that bolster the reliability of science and that are wholly compatible with value-laden science. We can have objective and value-laden science, and explicating how this is possible clarifies the basis for science's reliability and authority.

Returning to the nuts and bolts of science in policymaking, chapter 7 concerns how we should understand the needed integrity for science in the policy process given the pervasive role for values in science. I will argue that attempts to separate science from policy have failed, but that the integrity of science can be brought into focus and defended in light of the philosophical work of the previous chapters. With a more precise view of scientific integrity, we can more readily understand the sound science–junk science debates and see our way through them.

Finally, in chapter 8 I present some examples of how these considerations lead to a different understanding of science in policy, and how that understanding includes an important role for the public in the practice of policymaking. In particular, I will address the problem of how to make the proper role of values in science accountable to the public in a democracy.

Philosophers might immediately object to the trajectory of this argument on the grounds that I am confusing the norms of theoretical and practical reason. In philosophy, this distinction divides the norms that should govern belief (theoretical) and those that should govern action (practical). The basic distinction between these realms is that values should not dictate our empirical beliefs (because desiring something to be true does not make it so), even as values might properly dictate our actions (because desiring something is a good reason to pursue a course of action). John Heil (1983, 1992) and Thomas Kelly (2002), for example, have challenged the sharpness of this distinction, and while I will draw from some of their work, I will not attempt to resolve the general tensions between theoretical and practical reason here. Instead, I will argue that (1) simply because science informs our empirical beliefs does not mean that when scientists make claims based on their work they are not performing actions; and (2) the intuition that values should *not* dictate beliefs is still sound. Indeed, I will argue that values dictating belief would violate the norms of good scientific reasoning, thus preserving an essential aspect of the distinction between theoretical and practical reason. But dictating beliefs is not the sole role values can play. And the actions of scientists as voices of authority cannot be handled properly by merely concerning ourselves with theoretical reasoning. Making claims is performing an action, and some concerns of practical reason must be addressed. How to do this without violating the core norms of theoretical reason is at the heart of this book.

The arguments I will present in the following chapters have been developed against the backdrop of current discussions in philosophy of science, particularly on values in science. In addition, there has been some philosophical attention to science in public policy since 1990, although this has not been a central area of concern (for reasons discussed in chapter 3). Before embarking on the trajectory I have laid out above, it will be helpful to situate the arguments to come among this work.

The most careful examiner and defender of the value-free ideal for science since the 1990s is probably Hugh Lacey. In his 1999 book, *Is Science Value-Free?*, Lacey develops a three-part analysis of what it means to be value-free. He distinguishes among autonomy (the idea that the direction of science should be completely distinct from societal concerns), neutrality (the idea that the results of science have no implications for our values), and impartiality (the idea that scientific reasoning in evaluating evidence should involve only cognitive and never social or ethical values) (Lacey 1999, chap. 10). Lacey strongly defends the impartiality thesis for science, arguing for a strict distinction between cognitive (for example, scope, simplicity, explanatory power) and noncognitive (for example, social or ethical) values, and for the exclusion of the latter from scientific reasoning. Lacey's conception of impartiality captures the current standard core of the value-free ideal, as we will see in chapter 3. He is more moderate in his defense of neutrality and autonomy, arguing that neutrality is only a plausible ideal if one has sufficiently diverse "strategies" or approaches to research within disciplines, something Lacey finds lacking in many areas of current scientific practice, particularly in the arena of plant genetics (Lacey 2005, 26–27). Autonomy in research is even more difficult to assure, as the importance of funding in science has grown (see chapter 2). And recent careful reflection on policymaking for science seems to suggest that autonomy may not be desirable in the ideal (see, for example, Guston 2000; Kitcher 2001).[9] While I will not address the issues of neutrality and autonomy here, I will be directly critiquing the ideal of impartiality, which Lacey views as logically prior to the other two. If my criticisms hold, then all three theses of value-free science must be rejected or replaced.

Hugh Lacey is not the only philosopher of science who has defended the value-free ideal for science while examining areas of science crucial for policymaking. Kristen Shrader-Frechette has held a steady focus on the role of science in policymaking, providing in-depth examinations of nuclear waste handling and concerns over toxic substances, and using these examples to develop concerns over the methodological flaws and weaknesses of some risk analysis processes (Shrader-Frechette 1991, 1993). Her views on the proper role for values in science have also followed the traditional value-free ideal. For example, in *Risk and Rationality* (1991), she argues that, "although complete freedom from value judgments cannot be achieved, it ought to be a *goal* or ideal of science and risk assessment" (44). In *Burying Uncertainty* (1993), when describing "methodological value judgments," she considers the traditional epistemic values, which are acceptable under the value-free ideal for science, and problems of interpretation with them (27–38). She explicates clearly how the reliance on these values can create problems in risk assessment, but no alternative norms for scientific reasoning are developed. It might seem she undermines the value-free ideal in her book *Ethics of Scientific Research* when she writes, "Although researchers can avoid allowing bias and cultural values to affect their work, methodological or epistemic values are never avoidable, in any research, because all scientists must use value judgments to deal with research situations involving incomplete data or methods" (Shrader-Frechette 1994, 53). However, the importance of the value-free ideal becomes apparent when Shrader-Frechette equates objectivity with keeping the influence of all values to a minimum, and still only "methodological" (or epistemic/cognitive) values are acceptable (ibid., 53). In this book, I will disagree with Shrader-Frechette on this point, arguing that the value-free ideal needs to be rejected *as an ideal*, and making a case for a replacement set of norms for scientific reasoning. In addition, Shrader-Frechette contends that scientists are obligated to consider the consequences of their work because of a professional duty as scientists (Shrader-Frechette 1994, chap. 2). I, however, will argue in chapter 4 that the obligation to consider the consequences of one's choices is not a duty special to a profession or role, but a duty all humans share.

The work of Helen Longino is probably the closest to my own position on how to understand the proper role of values in science. Rather than starting with a focus on policymaking, Longino has worked from the feminist philosophy of science literature that developed out of feminist critiques of science in the 1980s. In *Science as Social Knowledge* (1990), she lays out a framework for understanding the ways in which values can influence science, particularly through the adoption of background assumptions. She distinguishes between constitutive and contextual values in science, arguing that both influence science in practice and content (4). Longino develops her account of values in science by examining the functioning of background assumptions in scientific research relating to gender. To address the concerns about the objectivity of science raised by these examples, she suggests that we think of science as an essentially social process, and she develops a socially based view of objectivity that can inform how science should function (ibid., chap. 4).[10] While her work serves as a useful starting point for more in-depth discussions on the role of values in science and the social nature of science, I depart from her framework in several ways. First, I want to provide a more closely argued account for how the adoption of an ostensibly empirical background assumption could "encode" values, work I have begun elsewhere (Douglas 2000, 2003a). Second, I do not utilize her distinction between contextual and constitutive values in science because I want to maintain a focus on both the scientific community and the broader community within which science functions, and dividing values into the internal and external at the start obscures one's vision at the boundary. In addition, as Longino herself argues, the distinction can provide no ground for ideals for science, as it is a thoroughly porous boundary between the two types of values. (This issue is discussed further in chapter 5.) Finally, while I appreciate and utilize Longino's emphasis on social aspects of science, I think we also need clear norms for *individual* reasoning in science, and this book aims to provide those. Thus, while my views on objectivity, as articulated in chapter 6, draw insight from Longino, I do not rest the nature of objectivity on social processes alone.

In addition to these philosophers who have grappled with the role of values in science, two writers have provided important insights on the role of science in policymaking. I see this book as expanding on the insights from these earlier works. Sheldon Krimsky, for example, has contributed much to the discussion on science and technology in public life and policymaking, focusing on the biological sciences and their import (Krimsky 1982, 1991, 2000, 2003; Krimsky and Wrubel 1998). Krimsky's discussions are wide ranging, and only some pick up on the themes of this book, as most of his work centers on the relationship between the public and the uses of new technology. The book that most closely relates to the concerns to be considered here is *Hormonal Chaos*, an overview of the endocrine disruptor debate, where Krimsky addresses problems of the acceptance of evidence as a basis for both science and policy (Krimsky 2000). While his discussion of that particular debate is rich, the general implications for understanding science in policy are not fully developed. For example, in order to get us beyond the sound science versus junk science debate, Krimsky briefly mentions a new ideal, "honest science," which he describes as "science that discloses financial interests and other social biases that may diminish the appearance of objectivity in the work" (ibid., 187). While this sketch is suggestive, it needs further development. Which interests are relevant and why? Why is the exposure of interests important to the integrity of science? How does this fit with the ideal of value-free science, in which one's interests are not to interfere with the interpretation of evidence? Answering these questions with more in-depth normative work is one of the purposes of this book. I will propose that it is not a full disclosure of interests that is needed for the integrity and objectivity of science, but an explicit and proper use of values in scientific reasoning. Not all interests are relevant to the doing of science, and some kinds of influence arising from interests are unacceptable, even if disclosed. Situating the scientist with respect to his or her interests is a good start, but not normatively sufficient.

In contrast to the work of Krimsky, Carl Cranor's *Regulating Toxic Sub-*

stances (1993) is more focused on the topic of the general use of science in public policy. Cranor examines the implications of accepting (or rejecting) certain levels of uncertainty in science to be used as a basis for policy, and provides a careful account of the processes of risk assessment and the uncertainties involved. I take up his focus on the trade-offs between underregulation and overregulation and expand their reach beyond how public officials and administrators should think about science to how scientists and philosophers of science should think about science, given science's central importance in the policy process. In particular, I will address how these insights lead to rethinking our understanding of norms for scientific reasoning, the nature of objectivity, and how to differentiate junk science from sound science.

Scientists and philosophers still largely hold to the value-free ideal.[11] Some claim that the value-free ideal is essential to the authority of science, to objectivity, or to the very possibility of having reliable knowledge (for example, Lacey 1999, 223). The critiques of the value-free ideal to date have been based on its untenability rather than its undesirability. I will take the latter road here and provide an alternative ideal in its place. Having a clearer understanding of how values should, and should not, play a role in science, working from the foundations of both moral responsibility and proper reasoning, should provide a clearer framework with which to examine the role of science in policymaking.

Thus, this book is about how scientists, once engaged in a particular area of research, should think about the evidence, and should present their findings, given the importance of science in policymaking. This area has reached a philosophical impasse of sorts. The value-free ideal requires that ethical and social values have no influence on scientific reasoning in the interpretation of evidence. But works like Longino's and Cranor's suggest that something is quite amiss with this ideal, that values and interests are influential for scientists, and perhaps properly so. What we need is a reexamination of the old ideal and a replacement of it with a new one. The ideal I propose here is not just for the social functioning of science as a community, but for the reasoning processes of individual scientists, for the practice of science and for the giving of scientific advice.

In addition to this philosophical literature on values in science and science in policy, there are related bodies of work that will not be directly addressed in this book. For example, Shrader-Frechette (1991, 1993), as well as Douglas MacLean (1986), have also done considerable work on which values shape, or should shape, our management of risk. This work argues for more ethically informed weighing of the consequences of policy decisions, suggesting that qualitative aspects of risk, such as its distribution, the right to avoid certain risks, voluntariness, and the valuation of life, are crucial to a complete understanding of risk in our society.[12] In this book, I do not engage in debates over which values in particular should inform our judgments concerning risk, but instead focus on how values in general should play a role in the science that informs our understanding of risk.

To finish setting the bounds of this book, a few caveats are in order. First, some hot-button issues will not be discussed. For example, debates over science education have been chronically prominent in the past few decades, as first creationism, and now intelligent design, seek to challenge the content of science education through school boards and textbook disputes, rather than through the scientific debate process of journal publications and conferences. My concerns here are not with what science to teach to the young, but with what science to depend upon to make decisions in public policy.[13]

Second, the book focuses exclusively on the natural sciences as a source of desired expertise. My neglect of the social sciences, such as psychology, sociology, and economics, arises partly from the need for less complexity in the book, and partly from a desire to avoid the debates over the scientific standing of the social sciences. Social sciences also raise unique problems of reflexivity, as the subjects of the research can read and understand the research, and alter their behavior as a result. How the ideas I develop here would apply to such contexts awaits future work.

Finally, while my treatment brings philosophical attention and analysis to the role of science in public policy, an issue much neglected by philosophers in the past forty years, science is a changing institution in modern society, and many have noted that the most important issues of science in society may not center on science and *public* policy as we enter the twenty-first century. Since the 1990s, the sources of funding for science and the institutional settings for research have been changing, with potentially important consequences (Greenberg 2001; Krimsky 2003). Over half of the funding for scientific research in the 2000s comes from private sources. With intellectual property concerns keeping much private research private, one may wonder whether the issues I address here are really salient. Such a critic would have a point, but until philosophy of science ceases to define itself solely in epistemological terms, such issues can hardly be addressed.

I see this book as part of a reorientation of the discipline of philosophy of science, to begin to take seriously again a philosophical (that is, conceptual and normative) examination of science as it functions in society in *all* of its aspects. With such a reorientation, we will be better able to address the issues presented by these changes in scientific practice, and the implications for policy for science, broadly construed.

CHAPTER 2

THE RISE OF THE SCIENCE ADVISOR

WITH THE CURRENT OMNIPRESENT NEED for science advice, how to ensure the soundness of such advice has become an ongoing source of difficulty in government. Yet the need for sound science advice was not always obvious. At the beginning of the twentieth century in the United States, there were no regular avenues for science advice to the government, much less regular contact between scientists and policymakers. Although the National Academy of Sciences had been founded during the Civil War to provide science advice to the government, after the Civil War the academy drifted from advising prominence to being primarily an honor society for the nation's budding scientific community. By 1900, there were some scientists embedded within governmental agencies with particular needs, but the larger scientific community had no relationship with the government. This pervasive absence of academic scientists in the halls of power was not a source of concern for policymakers, politicians, or academic scientists.

A century later, the relationship between science and government has blossomed. Even with the chronic debates over which science is most trustworthy, the need for regular science advice is unquestioned. In the United States, federal laws and federal regulations help to structure and ensure the quality of scientific advice, and many avenues and standing bodies provide such advice. Although there are debates over whether science advisors are heeded too much or too little, the need for science advice from a broad base of scientists is obvious. How did this dramatic shift in the perceived importance of science advice come about? In this chapter, I will chronicle the rise of the science advisor in the United States, from an occasional partner in governmental endeavors to a mainstay of the policy process.[1]

Several aspects of this story are of particular note. First, World War II is not the beginning of the story, as is often assumed. Crucial tentative steps and experimentation with how to develop a science-government relationship took place prior to World War II. Second, the successes from the pre–World War II era served as the basis for the rapid scaling up that takes place post-1940. This should give hope to any who wonder whether new institutional mechanisms for grappling with science advising in a democracy can take hold (mechanisms which will be explored in chapter 8). Rapid expansion can occur, even as institutions remain fluid. Third, the scientist-advisor becomes a pervasive feature by the late 1950s, the same time that philosophers of science come to ignore this important role (as we will see in chapter 3). Finally, the tensions that would drive the sound science–junk science debates discussed in the last chapter are apparent by the 1970s, even as earlier problems with science advising are resolved.

The rise of the science advisor was not a simple, linear path. Prior to World War II, science advising in the United States occurred in fits and starts, producing uneven success. World War II was the watershed event that made permanent a close relationship between science and government. Once the need for a strong relationship was cemented, the question remained what shape that relationship should take. The success of advising during the war depended on close, personal ties that could not be sustained in the context of the burgeoning science and government relationship after World War II. Although the war demonstrated the pervasive importance of science to government, it did not establish how the relationship between science and government should be structured. Further experimentation and evolution in that relationship took place as the institutionalization of the relationship deepened. Key templates were drawn, however, from the first half of the twentieth century, most notably the national lab, the contract grant, and the science advisory committee.

Science and the U.S. Government Prior to 1940

A relationship between science and government in the United States was not created wholesale from the crucible of World War II. The makings of that relationship had roots at least as deep as the nineteenth century, such that by 1900 the importance of having some scientists in government was clear. As Dupree (1957) wrote in his chronicles of science and the federal government, "In the first years of the twentieth century a government without science was already unthinkable" (288). The science that was a part of many government activities was located in a scattered federal research establishment, with agencies as diverse as the Geological Survey, the National Bureau of Standards, the Department of Agriculture, and the Bureau of the Census.[2] In the first decade of the twentieth century, the Public Health Service was added to this pantheon. These diverse bureaucratic entities engaged in scientific research to help guide policy decisions and to assist with scientific issues of public interest. Thus, having some needed scientific activity within the federal government was a given, but the relationship was generally ad hoc and internal, generating scientific research within the government when and where it was deemed necessary. Ties to the broader scientific community were largely absent.

In addition to this budding federal research establishment, there was a little utilized mechanism for seeking external scientific advice. In 1900, the sole formal science advisory body in existence was the National Academy of Sciences (NAS), created as a separate entity from the U.S. government during the Civil War to provide a standing resource for external science advice. According to its charter, the NAS was to respond "whenever called upon by any Department of the Government" to requests to investigate scientific issues. The NAS was to have its expenses reimbursed for such work, but was to receive no compensation (Dupree 1957, 139–40); thus was the independence of the academy to be assured. However, by the end of the nineteenth century, the inherent passivity embedded in the charter—to not respond unless called upon—had turned the NAS into a standing honor society, not an active advisor to the government (ibid., 294).

The compartmentalized nature of federal research and the inactivity of the NAS created new challenges with the coming of World War I. In fact, when the need for outside expertise became apparent after the sinking of the *Lusitania* in 1915, it was not to the NAS that the government first turned. Instead, Secretary of the Navy Josephus Daniels contacted Thomas Edison (who, as an engineer, was not considered a proper candidate for the NAS) to help organize a committee of experts to screen the numerous unsolicited inventions that were being sent to the navy to aid with the coming war effort (Dupree 1957, 306; Kevles 1995, 105). In addition to the unpaid screening efforts of the newly formed Naval Consulting Board (NCB), Edison and Daniels persuaded Congress that monies were needed to fund research directed by the NCB, research that could be more properly targeted at the needs of the navy than the whims of America's many inventors. The $1 million appropriation was the beginning of the Naval Research Laboratory (Dupree 1957, 307).[3] National Academy members were nowhere to be seen on the NCB, which was composed primarily of engineers. When asked why the NAS was so absent from the NCB, one inventor quipped, "Possibly because they have not been sufficiently active to impress their existence on Mr. Edison's mind" (quoted in Kevles 1995, 109).

This state of affairs aggravated at least one NAS member, astronomer George Ellery Hale, who began pushing for more NAS involvement with the war effort. The National Academy proved too stultified to be of assistance to the government, and the National Research Council (NRC) was formed in 1916 as a more active arm of the NAS (Dupree 1957, 309; Kevles 1995, 112). By 1917, the NRC and the NCB were in direct competition over which body could produce a reliable submarine detector first (Kevles 1995, 116). Both groups had some success developing early sonar devices, although scholars dispute whether these early devices or the convoy system was more important in protecting shipping from submarine attacks.[4] What *was* successful was the mobilization of a large-scale research effort targeted at a particular problem. The NRC was made a permanent body in May 1918.

As World War I ended, the NRC sank into inactivity with respect to the government, instead focusing on helping to bring U.S. scientists from various scientific societies together (Dupree 1957, 329). The growth in science funding and institutions for the next decade took place largely in the booming industrial labs of the period. The one major effort to provide support for basic research in the United States, the National Research Fund proposed by Hoover in 1926, was supposed to draw contributions from the private sector, but it was a "complete and immediate failure" (ibid., 342). Against this backdrop of government malaise with respect to science stands the bright spot of the National Advisory Committee on Aeronautics (NACA). Formed in 1915 to help the United States keep pace with developments in the new arena of flight, NACA was composed of twelve presidential appointees, with members drawn from the army, navy, Weather Bureau, Bureau of Standards, the Smithsonian, and the public—that is, nongovernmental scientists with relevant expertise (ibid., 287). By 1920, NACA had its own lab at Langley Field, which included one of the world's few wind tunnels (Zachary 1997, 85). The National Advisory Committee on Aeronautics also collaborated with universities in conducting their research (Dupree 1957, 334).

In the 1920s, NACA and its lab thrived and its annual budget grew from $5,000 in 1915 to $500,000 by 1927 (Kevles 1995, 105; Dupree 1957, 334).[5] The advisory committee provided a successful model for how government could support scientific research, attract good outside advice, and remain flexible in the face of change.

With the onset of the Great Depression in 1929, scientific research suffered. Budgets for science dropped precipitously by 1933 throughout government, as well as in industrial labs and universities. Indeed, some believed that science had helped to cause the Depression by making production so efficient that workers were no longer needed. Even among the many who still valued science, there was little impetus to fund science, a provider of long-term societal benefits when acute crises of employment and hunger were at hand (Dupree 1957, 345–47). Despite the decline of funding for science, the apparent need for good science advice grew during this period, particularly with a New Deal emphasis on careful planning and use of resources. With the NRC and the NAS largely quiescent, when a request came from Agriculture Secretary Henry A. Wallace for advice on the reorganization of the Weather Bureau, the NRC replied that it could provide little help with the task. Instead, it suggested that a more flexible policy-oriented body be formed within the auspices of the NAS. By executive order, President Roosevelt created the Science Advisory Board (SAB) in 1933, a committee of prominent scientists in academia and industry who would create and run subcommittees focused on all manner of scientific issues important to government. Massachusetts Institute of Technology President Karl Compton was named chairman of the SAB, with other board members drawn largely from the natural scientists who were part of NAS. Within two years, over one hundred scientists had served on SAB subcommittees, and numerous reports and recommendations had been made to the government on the organization of research and other technical questions (Compton 1936, 35–36).

Yet the successes of the SAB in providing advice to the government's scientific bureaucracy were overshadowed by two prominent failures. First, the SAB pushed for funding to employ out-of-work scientists in research endeavors for the public interest. The proposal was ultimately rejected, much to Compton's disappointment (Dupree 1957, 353–57; Pursell 1965, 346). Second, internal divisions arising in 1934 over the appointment of non–academy members to the SAB undermined its ability to be effective (Pursell 1965, 347–48). When the more broadly based National Resources Board (NRB) began to look at the need for science research planning, the apparent need for the SAB abated. The SAB was abolished by the end of 1935, and its planned successor body in the NAS rapidly faded into oblivion (Dupree 1957, 358).

Even though the SAB failed to provide the desired permanent avenue for science advice, it did presage many of the issues that would arise in the coming decades. First, even as it faded from the scene, the SAB predicted that the need for science advice would only grow. As noted in its second annual report,

> With government in a stage of transition from the more passive and regulatory part played in the past, to one of more intelligent and broad supervision and initiation, it is the concern of every citizen that there be available to government the most competent and impartial advice which can be found. The endurance of our traditional form of government will depend in increasing measure upon the quality of expert judgment, tempered with experience, which is available to government, and the willingness of government to follow such judgment. (Quoted in Compton 1936, 31)

How to obtain trustworthy advice would be an issue throughout the coming decades, as would how to keep science advisors from being abused. Indeed, Compton noted the presence of tensions still remarked upon today in the science advising process: "As fact-finding agencies the scientific services should be free to produce results that are not discolored by the opinions and snap judgments of policy-making political groups who may wish to put the dignity of 'science' behind their plans in order to win public approval" (ibid., 32).

An additional important shift in the relationship between science and government arose from the work of the NRB, altering the landscape of how science was viewed by the government.[6] While the SAB had drawn its ranks almost exclusively from natural scientists, the NRB included social scientists and educators as well, thus bringing together a more comprehensive view of research activities and planning possibilities. Interested in understanding and assisting all of these intellectual enterprises together, the NRB came to view research as a national resource. It broadened concern for research beyond government scientists focused on particular problems to include scientists in industry and academia, and suggested that all of these research activities were a national concern and thus collectively of interest to the government. As noted by Dupree (1957), this was a major shift in the view of science: "The logic of the broadened view of research as a national resource led to a concept of the government's responsibility extending beyond its own [science] establishment" (360). By the late 1930s, the idea that government should fund science beyond the activities of government scientists was being put into practice by pioneering agencies such as the National Cancer Institute (created in 1937) and NACA (ibid., 366). And one key way in which these agencies provided external science funding was through the contract research grant, which provided funds to scientists for particular research projects. This allowed for both autonomy for the scientist and accountability to the agency providing the funding. It was a mechanism that addressed many of the concerns over science funding between scientists and the government, one whose use would expand greatly in the coming decades.

As World War II approached, the SAB had been dissolved, with no mechanism for general science advising replacing it. Despite the SAB's call for the importance of science advice, new mechanisms and pathways for advising would have to be developed in the crisis of World War II. But the decades leading up to the war provided much of the raw material with which the science-government relationship would be so greatly expanded, from national labs to advisory bodies to contract research grants.

The Watershed of the Second World War

World War II, the largest conflict in human history, initiated a new relationship between science and government in the United States. Central to this change was the fact that the importance of science for American survival and prosperity were amply illustrated during the war. The stunning successes of radar, penicillin, and most dramatically, the atomic bomb, made the country aware of how powerful an ally science could be. This obvious importance would translate into permanent government funding sources for nongovernment scientists in the post–World War II era. In addition, the need to have scientists close to government decisionmakers also became apparent. Without technical men in the halls of power, many of the science-based successes of the war would not have come to pass, as the effort would not have been expended to bring those endeavors to fruition. Neither the funding of nongovernment scientists nor the use of external science advisors was entirely new, but the scale and intensity of such interactions vaulted these relationships to a prominence not seen before.

What was needed to unite and grow the scattered endeavors of the 1920s and 1930s into an all-out war effort was the right person in the right position. That person was Vannevar Bush, who would become President Roosevelt's personal liaison to the science community. Made president of the Carnegie Institution of Washington in January 1939, Bush had previous experience in both industrial science (as a director of the company Raytheon) and academic science (as professor of electrical engineering and subsequently vice-president of MIT) (Zachary 1997, 39–45, 84). In addition, by October 1939 he was chairman of NACA, and thus understood both its functioning and its successes (ibid., 99; Dupree 1957, 370). Bush embodied the experience needed to organize a war effort across the arenas of science in the United States. Although Bush was a Yankee Republican, Democrat Roosevelt found in Bush the link he needed to the scientific community. When Bush met personally with Roosevelt in June 1940 about forming a new organization to mobilize the full scientific community in the United States for the coming war effort, Roosevelt quickly signed off on everything for which Bush asked (Zachary 1997, 112; Kevles 1995, 297). As summarized by Bush's biographer, "In minutes, the president had promised Bush a direct line to the White House, virtual immunity from congressional oversight, and his own line of funds. The seeds had been sown for their extraordinary relationship" (Zachary 1997, 112). Solely by Roosevelt's authority was the National Defense Research Committee (NDRC) formed, operating entirely out of presidential discretionary funds, and thus not requiring any congressional oversight.

The members of the NDRC were drawn from the government departments of war, navy, and commerce, plus members from the National Academy of Sciences and other esteemed scientists, with Bush as the chairman.[7] Bush recruited James Conant (president of Harvard), Karl Compton (president of MIT and former head of the Depression-era SAB), and Frank Jewett (president of the NAS and head of AT&T's Bell Laboratories) to help him run the fledgling council. The crucial task of the NDRC was to structure the giving of government support to nongovernment scientists in a way that would not threaten the autonomy so valued by the scientists (Greenberg 1999, 79; Zachary 1997, 114–16). Bush adapted the contract research grant approach from NACA for wartime, and began the dispersal of funds on a much larger scale than ever seen before. The formal framework of the contract research grant, within which science and government could work closely together and still tolerate each other, changed the traditionally frosty relationship between the scientific community and government officials. With contract research grants, the government could pay scientists who were not government employees to conduct research, which allowed university scientists to do crucial pieces of work for the government without forcing the scientists to leave the university setting. In the first year of NDRC's existence, it had a budget of over $6 million and authorized over one hundred research contracts to universities and industrial labs (Zachary 1997, 117; Kevles 1995, 298).

Although large compared to previous grant-funding operations, within a year it was apparent to Bush that the NDRC would not be sufficient for the war effort. By the spring of 1941, even before the United States formally entered the war, the research projects supported by the NDRC had grown, and mere research by the nation's top scientists no longer seemed adequate to meet pressing defense needs. Development of research into usable assets was also needed, which would require substantially more money. In response, Bush helped create the Office of Scientific Research and Development (OSRD) in May 1941, which subsumed the yearling NDRC (Kevles 1995, 299–300; Zachary 1997, 129). The OSRD received its appropriations from Congress, allowing much larger sums to be funneled through the office, sums adequate to develop microwave radar, proximity fuses, solid fuel rockets, and penicillin into usable tools for the military.

This expansion of research effort occurred *before* the United States committed to the largest of all scientific projects of the war, the Manhattan Project. It was not until the fall of 1941 that the crucial report from Britain arrived in the states with the analysis that made the atom bomb project look feasible. Prior to this, there were deep concerns over the technical possibility of generating sufficient fissile material for a weapon, as it was thought both that more fissile material was needed than is the case and that the difficulties in generating the material were likely insurmountable. The British report (known as the MAUD report[8]) made it clear how feasible producing an atom bomb was, thus prompting a serious commitment to the project in late November 1941, just before Pearl Harbor. By mid-1942, the Manhattan Project had been taken over by the military, and large-scale national labs were being planned to complete the project, with military security. (The army took over the directing of construction for the project sites— Oak Ridge, Hanford, and Los Alamos—and the majority of the funding) (Hewlett and Anderson 1962, 73–74). Thus, the OSRD's scale of effort had little to do with the Manhattan Project, and speaks to the enormity of all the other research efforts occurring during the war. Indeed, the costs for the radar systems alone ($1.5 billion) rivaled the costs for the entire atom bomb project ($2 billion) (Kevles 1995, 308). For many of the physicists involved in these projects, the bomb project was not the central activity of the war.

As Kevles notes, "The atom bomb only ended the war. Radar won it" (ibid., 308).

In addition to the research efforts set up by OSRD at existing universities, new large national labs were created at Argonne, then Oak Ridge and Los Alamos. Although the idea of a national laboratory had a long history in the United States, dating back at least to the founding of the Naval Observatory in 1842, and continued in such efforts as NACA's lab at Langley Field, the size of the labs originated in the 1940s tends to obliterate their predecessors from view (see Westwick 2003, 27). The earlier, smaller labs served as inspiration for the World War II labs, although the national lab idea had never been implemented on so grand a scale.

The scientific endeavors during World War II also changed how scientists viewed both the government and the process of research. Scientists found having such grand efforts, and the resources involved, a wonderful experience (Greenberg 1999, 98–99), and they also enjoyed having access to the halls of government. But the relationship between Bush and Roosevelt was primarily a personal one. No formal systems that could continue it beyond wartime, and outside of the trusting relationship between Bush and Roosevelt, had been established. Although the need for having men of science in close contact with men of power seemed clear, how this was to work beyond the scope of the war was not.

More obvious at the close of the war was the need for continued federal funding of science. Only Frank Jewett argued against some continuation of the new, pervasive, and close ties between science and government. Jewett suggested in 1945 that government funding of science was not needed because the "amounts of money involved in support of first-class fundamental science research—the only kind that is worthwhile—are infinitesimally small in comparison with national income," and thus could be supported by private sources (quoted in Greenberg 1999, 141). However, as Greenberg notes, "Jewett was a minority of one" on this issue (141). The rest of American scientists enjoyed the possibilities that large-scale federal funding presented to them, and political leaders, shaped in part by Bush's *Science: The Endless Frontier*, desired to make continued use of the security and largesse science could provide the country.

Thus, it was generally accepted at the end of the war that science and government would have to maintain the close ties developed during the war. However, how these ties were to be institutionalized was not clear. With a sense of urgency, but not a definitive plan, decentralization of the science-government relationship characterized the immediate postwar period. But unlike the previous crises of the Civil War and World War I, where close affiliations (or attempts at institutionalizing a link) between science and government faded and drifted after the wars' end, the need for the ties went unquestioned and thus eventually the ties became stronger after World War II. Although there were debates over the form of that relationship in the postwar years, the general effort to find acceptable institutional avenues between science and government never abated. For this reason, the Second World War is seen as a watershed in the history of American science policy.[9]

The Ascent and Institutionalization of Science Advisors (1945–70)

As World War II drew to a close in 1945, and Bush published *Science: The Endless Frontier*, it became clear that new administrative structures were needed between science and government. The personal relationship Bush had enjoyed with the president had died with Roosevelt that April. Truman had much less interest in or tolerance for the scientists who ran the OSRD. The organization could not survive the war, in any event, having been premised upon the extreme conditions of wartime, and built for speed and secrecy. New organizations would be needed for the postwar era. But the debates in the years immediately after the war over the details of the new National Science Foundation and over the control of atomic energy consumed the efforts of scientists and policymakers. These debates would shape the science policy landscape for decades, but prevented any efforts at creating a comprehensive approach to policy for science, or to science advice for policy. Instead, approaches and organizations were developed as needed, leading to the pluralistic, but pervasive, system we have today.

The debate over the founding of the National Science Foundation has been elaborated in detail elsewhere (see, for example, Kleinman 1995; Bronk 1975). Although there was little disagreement on the need for a federal agency that dispersed contract grants to scientists throughout the country, disagreements did arise over the precise shape of such an agency. Questions such as whether geographic distribution should be a consideration in funding decisions, whether the social sciences should be included in the NSF funding stream, and most importantly, the amount of control the president should have over the NSF, thwarted a quick resolution on legislation. Indeed, the issue of presidential control most delayed the creation of the NSF, preventing final legislative passage until 1950. Truman refused to sign any legislation that did not place the head of the NSF directly under his authority (as opposed to under the authority of a board of trustees).

While this debate delayed the creation of the NSF, other avenues of funding for science sprang up to amply fill the void—particularly through the newly created Office of Naval Research (ONR) and Atomic Energy Commission (AEC).[10]

As possible avenues for research funding proliferated (including NSF, ONR, AEC, NACA, and the National Institutes of Health [NIH] by 1951), creating a patchwork of funding sources rather than a single unified source, the avenues for science advising also proliferated, albeit at a slower pace. Bush, lacking good personal ties to Truman, felt no longer needed by the White House, and left in early 1946 (Zachary 1997, 302–9). No science advisor took his place; instead, science advising moved to the agency level. The most prominent of the time was the General Advisory Committee (GAC) to the AEC. The creation of the AEC was perhaps the most important political success of the scientific community after World War II, as it ensured civilian (rather than military) control over nuclear power.[11] The Atomic Energy Act of 1946, which created the AEC, also mandated the creation of the GAC to provide advice to the AEC. Chaired by Robert Oppenheimer from 1946 to 1952, and then by I. I. Rabi from 1952 to 1956, the GAC boasted prominent scientists such as Glenn Seaborg, Lee DuBridge, and Enrico Fermi as members. Although the GAC's opposition to the hydrogen bomb was not heeded (see Herken 2000, chap. 3), it had a powerful influence over policy, particularly in its early years. As Hewlett and Duncan (1962) described the committee, with its "distinguished membership," "it spoke with the voice of authority" (46). The first chair of the AEC, David E. Lilienthal, "fell into the habit of asking the GAC for all kinds of advice" (Lapp 1965, 104). While the GAC was the primary advising body to the AEC, the AEC also developed additional science advisory bodies, such as the Advisory Committee on Reactor Safeguards, as needed (Brooks 1964, 75).

In addition to the AEC's GAC, other advising committees of scientists also popped up in the post–World War II era. Most concerned themselves with advice on the direction of science funding. One early example, instigated by Bush's efforts to generate continuous science advising for the direction of military scientific research, was the Joint Research and Development Board, created in 1946. The board oversaw an elaborate collection of advisory committees, drawn from scientists in civilian life (Price 1954, 144–45). While the board allowed scientists to maintain contact with the military in the postwar years, it ultimately proved unworkable, both because the military was unwilling to allow the board to have final decisionmaking authority over research projects and because the board was not structured well to serve the needs of the military (ibid., 145–52; Smith 1992, 49). The board was ultimately abolished in 1953, but the loss of science advice to the military as a whole would not last. In 1956, the Defense Science Board was formed, an institution that successfully served the military in the decades following (Smith 1992, 50–60).

Although scientists were experimenting with multiple ways to advise various parts of the government, the five years after World War II had failed to produce any clear avenue for science advice for the president. Truman did not create an official science advisory body in the White House until 1951. War was still an impetus for bringing scientists into closer contact; the Korean War, in this case, was the prime motivator for Truman's request to William Golden for a report on better ways to utilize civilian scientists (Herken 2000, 55). Golden recommended that Truman create the position of a personal science advisor to the president, with an attached advisory committee (ibid., 55; Smith 1990, 111–12). However, the governing board of the newly formed NSF was concerned over the possible existence of the presidential science advisor, and initially opposed the plan on the grounds that such an advisor would undercut the function of the NSF to formulate national science policy (Bronk 1974, 117). In response to these concerns among scientists, Truman created the Science Advisory Committee (SAC), consisting of eleven prominent scientists, but he placed it within the executive branch's Office of Defense Mobilization (ODM), not in direct contact with the president (Burger 1980, 7–8). Although SAC was supposed to be a source of advice for both the president and the ODM director, Truman rarely consulted it (Smith 1992, 163). In the meantime, the NSF did little to direct national science policy. With the pluralistic funding structure in place, and an explicit demand that the NSF focus on basic research only (avoiding the defense research controlled by the Department of Defense [DOD] and AEC and leaving medical research to the NIH), the NSF worked on setting up its own administrative procedures in the 1950s and abandoned efforts to direct general science policy (Walker 1967; Bronk 1975).

With the change of president in 1952, the role of scientists in government again shifted. Because of an inherent interest in science advice, Eisenhower made better use of the Office of Defense Mobilization's SAC, requesting their advice on issues such as intercontinental ballistic missiles and the alert status of U.S. bombers (Smith 1990, 112; Smith 1992, 164). The launch of Sputnik in 1957 caused Eisenhower to elevate SAC to the body initially envisioned by Golden six years earlier. Eisenhower renamed SAC the Presidential Science Advisory Committee (PSAC), and appointed

James Killian as his special assistant to the president for science and technology (Burger 1980, 7). The Presidential Science Advisory Committee then named Killian as chair. At last, the president had an official science advisor in the White House. Although PSAC would survive as an institution until the 1970s, its heyday was in its first six years. Both Eisenhower and Kennedy were interested in hearing from PSAC, and they listened conscientiously to their science advisor. Thus, 1957–63 is often called "the golden age of presidential science advising" (Smith 1992, 165).

Sputnik inspired increased interest in and concern about science from many quarters, not just the president.[12] Scientists organized the Parliament of Science, convened by the American Association for the Advancement of Science (AAAS) in March 1958 to address a range of issues relating to science and society, including the need for science advice in government. The report from the meeting concluded that "representation of science and technology at the highest levels of government where . . . national policies are formulated" was needed, and as the newly elevated PSAC and science advisor "does seem to furnish the necessary high-level representation," the report "strongly endorse[d] the continuation of such an arrangement" (Parliament of Science 1958, 855). Some observers saw the new PSAC and its chair as a crucial counterbalance to science advice from more militant scientists who had the ear of the AEC (see, for example, Lapp 1965, 135). By 1960, roughly one hundred scientists served in an advisory capacity to the government through PSAC panels (Gilpin 1964, 8n16). By 1963, that number had increased to 290, only 32 of whom were directly employed by the government (Leiserson 1965, 412). Thus, PSAC became a major avenue for civilian and university scientists' influence on federal policy.

Yet pressures for increased scientist involvement continued, even with the creation and expansion of PSAC. As the government-science institutional structure grew, PSAC was quickly overburdened. To alleviate the rapidly growing set of tasks before PSAC, the Federal Council for Science and Technology (FCST) was formed in 1959 to address issues arising from within the government research structure, such as tensions among NASA, the NSF, and the AEC (Wolfle 1959, 48–49; Leiserson 1965, 414). This allowed PSAC to focus on general policy advice and the concerns of the broader scientific community. To help coordinate the tasks of PSAC and FCST, the science advisor to the president served as chair of both. In 1962, President Kennedy added yet another science advising body, the Office of Science and Technology (OST), to be headed by the chair of PSAC and FCST (Leiserson 1965, 415). The OST was finally to provide a home for the general coordination of research, the job the NSF was never able to take on, as well as a liaison for Congress (Lapp 1965, 199–200). Jerome Wiesner, Kennedy's official science advisor, thus had three additional hats to wear: chair of PSAC, chair of FCST, and director of OST. Some called him the "science czar" (ibid., 200).

The need for science advice was felt not just at the level of the president. In addition to this elevation and centralization of science advice in the White House, there was a growing sense that one needed good science advice for policy development across the executive branch. For example, the State Department set up the Office of Science Advisor in the late 1950s to assist the department and its embassies on matters scientific (Wolfle 1959, 31). Beyond the federal government, attempts to set up science advisory bodies at the state and local level, in imitation of the federal government, proliferated in the 1960s, with uneven success (Sapolsky 1968). Even political parties saw the need for science advice. The Democratic National Committee appointed an Advisory Committee on Science and Technology, made up primarily of prominent university scientists from a wide range of fields, to advise its Democratic Advisory Council (Lapp 1965, 181–85); the committee existed to help devise policy platforms for the 1960 election. The importance of scientists providing advice became increasingly obvious across government institutions and was viewed as something requiring serious attention. The United States had come quite far from the days of the First World War, when those in government did not bother approaching scientists for advice.

With this rise in prominence came an increased reflection by scientists on the proper role of the science advisor. According to the scientists involved with the Democratic National Committee, their role was motivated by the idea "that the citizen-scientist has a responsibility to think about the problems of science and society and to communicate his thoughts to those who can convert ideas into the fabric of national policy" (ibid., 182). But the scientist was not to become the politician, as "scientific and technological facts should not be the property of any political party." Concerns over protecting the integrity of science, echoes from the 1930s era SAB, were becoming more prominent. But the primary focus remained on being available to serve society by giving advice. This ethos of providing advice led to public service both within the government proper and in the public eye in general.[13]

As the institutional structures for science advice proliferated in the late 1950s and early 1960s, the general number of scientists serving in advisory capacities and the time commitment of those scientists also increased greatly. In addition to the formal commitment to sit on a standing advisory body, many scientists participated in the less formal summer studies programs, in which scientists would gather to discuss a particular problem and to provide targeted advice (Greenberg 1999, 28). Rather than complaining about being ignored, as they had prior to World War II, some scientists grumbled about the time advising the government took away from their research. For example, in 1962 Hans Bethe noted that he "spent about half of [his] time on things that do not have any direct connection with my science," and worried about the impact of such extra-research commitments on both the senior statesmen of science like himself and on younger scientists (quoted in Wood 1964, 41–42).

Despite the apparent quandary of being called upon too much, difficult times lay ahead for science advisors. Part of the difficulty would arise from strained relations between PSAC and President Johnson (Herken 2000, 146–64). But personal ties were not the only thing at issue. Although the prominence of science advisors had been on the rise, deep tensions in the structure of science advising lay dormant, just below the surface. One of the most important was the extent to which science advisors to the president were free to speak their thoughts publicly, including in front of Congress. This was a particular problem for the special science advisor to the president, who was also head of OST. His advice to the president was confidential, but as director of OST, he had an obligation to be forthcoming with Congress about the state of U.S. research (Leiserson 1965, 415; Price 1965, 242). How this tension over confidentiality and openness was to be resolved remained unclear. In addition, as budgets tightened with the expense of the Vietnam War, Congress increasingly demanded tangible results from research expenditures, while constraining the expansion of research budgets. Finally, many members of the public had developed antiscience views by the end of the 1960s (Brooks 1968; Price 1969). All of these factors placed strain on the advising system, strain that would become explosive by the early 1970s.

Controversy and Regulation of Science Advisors (1965–75)

By the mid-1960s, the federal advisory structure was well established, if impermanent. However, tensions arose over the proper role of the science advisor, and where such an advisor's loyalties ultimately resided. These tensions would bring about the dissolution of PSAC in the early 1970s, but not the end of science advising. Instead, institutional structures for science advising again proliferated, and tensions that arose were mitigated (or even resolved) by the passage of the Federal Advisory Committee Act (FACA) in 1972. The need for this act indicates how central science advice for policy had become to the policy process. The importance of and difficulties with science advising warranted its own legislation.

The passage of FACA had its origins in the difficulties of PSAC under both the Johnson and Nixon administrations. As noted in the previous section, PSAC, with all of its subcommittees, involved several hundred scientists by the mid-1960s, and this wide base continued under the Nixon administration (Perl 1971, 1211). However, divisions within the scientific community undermined the effectiveness of PSAC. In particular, scientists supporting weapons test bans or opposing new armaments were often pitted against scientists supporting new weapons systems. In the 1960s, the "peace" scientists often advised the government through PSAC, whereas the "war" scientists often advised the government through the Department of Defense or the national laboratories (Smith 1992, 167). Johnson was irritated by scientists who spoke against the Vietnam War while serving in an advisory capacity (ibid., 168); Nixon was even less tolerant of criticism coming from science advisors.

Initially, it was hoped that the standing of the science advisor would improve with the change from the Johnson to the Nixon administration. But early in Nixon's first term, Lee DuBridge, Nixon's special assistant for science and technology, was frozen out of national security and budget decisions (ibid., 169–70). Although PSAC and DuBridge had other issues on which they could focus, particularly the public's increasing concern over environment and health issues, their effectiveness was uneven (ibid., 170). The downfall of PSAC began when the first of two scandals involving the advisory committee and the Nixon White House erupted in 1969. PSAC was not consulted directly on Nixon's antiballistic missile (ABM) program (having been excluded from many national security decisions); however, prominent PSAC alumni provided a steady stream of criticism in congressional testimony about the program. DuBridge, attempting to help Nixon, drafted a letter of support, but was unable to get other PSAC members to sign it (ibid., 172–73). These events did not sit well with the administration; as Bruce L. R. Smith recounts, "Nixon and Kissinger were angered at the lack of support from the scientists, and Kissinger was reinforced in his belief that the committee [PSAC] was filled with administration critics operating under the pretense of scientific neutrality." The scientists of PSAC, "who saw themselves as the essence of objectivity," viewed the episode as "proof that the administration did not want advice but merely public support" (ibid., 173). Administration officials, on the other hand, considered the behavior of the scientists to be disloyal, and believed that science advising "was becoming chaotic" (Henry Kissinger, quoted in ibid., 173).

The supersonic transport (SST) controversy, coming right on the heels of the ABM issue, exacerbated the rift between PSAC and the administration. An ad hoc PSAC panel, headed by Richard Garwin, had been put together to study supersonic transport. The Garwin report, finished in 1969, was highly critical of SST's prospects, for both economic and environmental reasons. The report was kept confidential and, in 1970, administration officials then testified in Congress that, "According to existing data and available evidence there is no evidence that SST operations will cause significant adverse effects on our atmosphere or our environment. That is the considered opinion of the scientific authorities who have counseled the government on these matters over the past five years" (quoted in Hippel and Primack 1972, 1167). This was in direct contradiction to the Garwin report, however. When Garwin was asked to testify before Congress on the issue, he made clear that the PSAC report was contrary to the administration position. Concerted effort by a member of Congress got the report released, and the administration was sorely embarrassed (Hippel and Primack 1972, 1167). After his reelection in 1972, Nixon asked for and received the resignations of all PSAC members and refused to refill the positions (Smith 1992, 175). He transferred the science advisor position to the National Science Foundation, out of the executive office. The advisory committee had been killed and the science advisor position would not be reinstated to the executive office until 1976, and then only by an act of Congress (ibid., 178).

The most obvious lesson of these controversies was the danger of confidential science advice given to the executive branch, particularly when dealing with issues not related to national security. The executive branch could either make public the detailed advisory report, or keep it confidential, and in either case make claims that the advisors supported the project or policy choice. Many observers saw this as a dangerous imbalance of power and/or an abuse of science's authority. Thus, Martin Perl lamented the limited advice scientists gave to Congress compared to the executive branch, and criticized the confidentiality of much science advice: "Large numbers of advisory reports are made public; but, unfortunately, it is just those reports which concern the most controversial and the most important technical questions that are often never made public, or only after a long delay. This is unfortunate . . . for the process of making technical decisions in a democracy" (Perl 1971, 1212–13). The problem of confidentiality led Frank von Hippel and Joel Primack (1972) to call for "public interest science," in which scientists would act to advise the public, Congress, or special interest groups directly in order to counter the authoritative but secretive weight of executive office science advice.

Congress, however, came up with a legislative response to the problem. After a lengthy development, the Federal Advisory Committee Act was made law in 1972. Among its provisions, which govern all federal advisory committees, are requirements that advisory committee meetings be publicly announced within a sufficient time frame prior to the meeting date, that meetings be open to the public (with few exceptions), that meeting minutes be made available to the public, and that committees be balanced in their composition (Jasanoff 1990, 46–48; Smith 1992, 25–31). Meetings were opened to the public to help prevent potential abuses of power. Advisory committee reports as well as deliberations were to be made public, except when national security dictated otherwise. Advice on public issues concerning the environment, consumer safety, or transportation choices could not be secret anymore. The advisory system would no longer be able to be used to create "a facade of prestige which tends to legitimize all technical decisions made by the President" (Perl 1971, 1214).

In addition to dealing with the problem of secret advice, FACA also addressed the second lesson from both the demise of PSAC and other prominent controversies of the day (such as the debates over DDT): that scientists can and would enter the political fray, often on opposing sides of an ostensibly technical issue. It was apparent that scientists could be experts and advocates simultaneously. (Indeed, Hippel and Primack called on scientists to be both advocates and experts.) If scientists could have such different views on technical topics that they naturally served as advocates for different sides, a balance of the range of scientific views was needed on advisory committees. The Federal Advisory Committee Act thus required that advisory committees be "fairly balanced in terms of points of view represented and functions to be performed" (FACA statute, quoted in Jasanoff 1990, 47). This requirement addresses the traditional concerns with getting the appropriate expertise for any given panel, that is, expertise from relevant disciplines, while also recognizing the range of scientific views one is likely to find among scientists and the need to balance them. Exactly what "balance" means, and which point of view should be considered, remains a source of controversy. Indeed, by the 1990s the problem of which scientific experts to trust (given the frequent presence of scientists on multiple sides of an issue) would blossom into the sound science–junk science debate discussed in chapter 1.

Despite these concerns over science advising and the role of scientists,[14] the need for scientists in advisory roles continued to grow in the 1970s, particularly with the expansion of the regulatory agencies. Although PSAC was dismantled by the executive office, this singular decline was more than made up for in the expansion of science advising at the agency level. Even in 1971, estimates of the number of scientists involved in the broadly construed executive advisory apparatus ran into the thousands (Perl 1971, 1212). New legislation (for example, the National Environmental Protection Act, Clean Air Act, and Clean Water Act) increased the need for science advice to inform public policy. New agencies were created to implement the regulatory laws (such as the Environmental Protection Agency [EPA], Occupational Safety and Health Administration [OSHA], and Consumer Product Safety Commission [CPSC]). To provide these agencies with independent (from their staff scientists) advice, science advisory boards were created and proliferated at an astonishing rate.[15]

While wielding no direct power over agencies (such power would be unconstitutional), these advisory boards have become a major tool for the legitimation of policy decisions. Often created initially by the agency in order to help vet their science-based policy decisions, these boards became mandated by Congress in many cases in order to ensure that science used in policymaking would be sound. A negative or highly critical review by an advisory board almost always sends a policy back to the agency for reconsideration.

However, as noted in chapter 1, these advisory boards have done little to squelch controversy over science-based policy. While it was hoped that the advisory boards might decrease the adversarial debates that were becoming the norm in science-based policymaking, they did not. Rather, they became yet another player embroiled in the debates.[16] The scientific uncertainties usually were too deep to be settled by adding another layer of scientific review. Even when multiple review panels came to the same conclusion, it was not clear that the scientific issues were settled.[17] In addition, the National Research Council's reports, considered the gold standard for science advice, often failed to produce clear consensus statements on key policy points, such as the risk saccharin poses to humans (Bazelon 1979, 278). Publicly dueling experts became the norm in high-profile policy decisions with scientific bases (see Nelkin 1975). Despite the now legislatively mandated and entrenched role for science advice in policymaking, chronic debates over sound science and junk science still occurred. The federal science advisory apparatus, now extensive enough to be called the "fifth branch of government," did not help to resolve these difficulties (Jasanoff 1990).

Conclusion

From the broad perspective of a century, the importance of science advising has steadily grown. Although there were fits and starts in developing the relationship between government and science advisors, the role of the science advisor changed from a sporadic, occasional influence, to a strong, personal relationship, to a deeply entrenched institutional necessity. Scientists now clearly occupy influential and important roles throughout our government, providing advice on a wide range of technically based issues. While the agencies that they advise cannot be required to follow their advice, it is uncomfortable and politically risky to ignore it, particularly since FACA was enacted, requiring that the advice be made public. In this regard, scientists giving advice wield real power and influence (even if their advice is not always followed as they would like). Even when science advice is divided, and we have competing or dueling experts, simply ignoring the experts is not a viable option. Yet on which experts to ultimately rely remains a source of contention. In this climate, disputing parties have been reduced to merely attempting to tarnish unwelcome expert opinion as "junk science" or "politicized science."

One might think that given the obvious normative tensions involved with science advising, and its importance in the second half of the twentieth century, philosophers of science might have been able to assist with sorting out the issues involved with science in public policy. Yet despite the clear authority of the scientist's voice in contemporary society, this aspect of science has been largely ignored by philosophers of science.[18] This neglect, which grew out of the crucible of the cold war, was key to the adoption of the current value-free ideal for science.[19]

CHAPTER 3

ORIGINS OF THE VALUE-FREE IDEAL FOR SCIENCE

WHILE SCIENTISTS TOOK ON an ever more visible, even if more contentious, public role throughout the 1960s and 1970s, philosophers of science came to ignore this public role. One might imagine that philosophers of science would illuminate this role, examining the place of expertise in a democracy and helping to shape public discussion of the proper relationship between science and society. Yet since the 1960s, philosophers of science have been largely silent on these issues. Most philosophers of science consider their work to belong to a subfield of epistemology, the study of knowledge, and as such are solely concerned with epistemological issues in science, such as the relationship between evidence and theory, the status of scientific theories, and the nature of scientific explanation. Issues of how to understand science in society, the role of social values in science, and the responsibilities of scientists have been excluded from the field.[1] Most of this exclusion has been maintained on the basis that science is (or should be) a value-free enterprise, and that scientists should not consider the broader implications of their work when conducting research. Under this view, there is nothing philosophically interesting about the relationship between science and society. Science is our best source for reliable knowledge about the world, and what society sees fit to do with that knowledge is its own affair, outside the purview of both scientists and philosophers of science.

Such a simple-minded understanding of science is not illuminating for the issues raised by science in policymaking. Nor have other discussions of science in society helped much. For example, while the Science Wars were precipitated by an attempt to directly challenge the epistemic authority of science, this debate was of little interest outside of academia and failed to provide insight on the complex role of science in society. That science should help answer certain questions and guide policy is not at issue. More central are the controversies concerning on which science (or scientists) we should rely, particularly given the protracted disagreements among scientific experts that exist in many cases. Society needs a more nuanced understanding of science in public policy, one that will account for how scientific experts can have lasting and intractable disagreements yet still be honest participants in a debate.[2] And both citizens and policymakers in a democracy must have a way to decide how to interpret scientific findings that are not settled science. A more careful appreciation of the role of values in science is essential to such an understanding. This approach, however, rejects the value-free ideal for science, going against the dominant position of the past forty years in philosophy of science.

In order to understand the value-free ideal in its current form, we need to examine in greater detail the origins of the value-free ideal for science. There are several surprises here. The first is that *in its present form,* the value-free ideal is fairly recent, dating predominantly from the cold war period.[3] In fact, it does not get its stranglehold on philosophy of science until around 1960. The ideal that has held sway since 1960 is a complex one. It does not hold that science is a completely value-free enterprise, acknowledging that social and ethical values help to direct the particular projects scientists undertake, and that scientists as humans cannot completely eliminate other value judgments. However, the value judgments internal to science, involving the evaluation and acceptance of scientific results at the heart of the research process, are to be as free as humanly possible of all social and ethical values. Those scientific judgments are to be driven by values wholly internal to the scientific community. Thus, the value-free ideal is more accurately the "internal scientific values only when performing scientific reasoning" ideal. How this ideal developed and became the dominant view in philosophy of science is chronicled in this chapter. I will also indicate how the philosophers' value-free ideal influenced the scientific community proper, and thus how it has hindered a useful and illuminating understanding of science in the broader scientific and policymaking communities.

The second surprise is that the ideal rests on a problematic presumption, one that is central to this inquiry. What philosophers of science needed to solidify the basis for the value-free ideal was the notion that scientists are not involved in public life, that they provide no crucial advisory functions, and that they provide no assistance for decisionmaking (or at least that scientists should act as if they were not involved with public life in these ways). In other words, philosophers of science assumed that science was fundamentally and acceptably isolated from society. Despite evidence to the contrary (as shown in chapter 2), philosophers continued to insist on this point into the 1980s, so it is not surprising that philosophy of science has been of so little assistance in helping to understand the role of science in public policy.[4] Once we reject the presumption of an isolated scientific community, the ideal of value-free science in its current form crumbles.

Philosophy of Science and the Value-Free Ideal

As Proctor (1991) discusses, the value-free ideal in science has meant different things in different times and places. Proctor locates the beginnings of the ideal in the seventeenth century with Francis Bacon and debates over the relationship between science and the state, but he argues that the ideal shifted in form and content in the nineteenth century, when the rise of the German university, the growing importance of the natural sciences, and the development of the social sciences led to stronger attempts to separate science from values (Proctor 1991, 65–74). As science became more important, the arguments for the value-neutrality of science, particularly for social science, found their strongest advocate in Max Weber. Yet this strong advocacy died with Weber in 1920 (ibid., 134, 157).

Thus, despite these roots in Western culture, the value-free ideal was *not* widely accepted by philosophers of science in the United States as recently as 1940.[5] This can be seen in an influential essay by Robert Merton, first published in 1942, which came to be known as "The Normative Structure of Science." The essay elaborates Merton's famous "ethos of science," with its four components of universalism, organized skepticism, communalism, and disinterestedness (Merton 1942, 270).[6] In the essay, Merton argues that democracies on the whole do a better job of supporting most aspects of the ethos of science than totalitarian states, a heartening argument in the darkest days of World War II. Two aspects of his argument are of particular importance: (1) that being "value-free" is nowhere among the norms, and (2) that the ethos is described as a set of norms internal to science but dependent on the broader society for their functioning. Of the four norms, the one that comes closest to a value-free ideal is disinterestedness. Merton, however, does not eschew a role for values in science in his discussion of disinterestedness, but instead argues that because of the institutional structure of science, scientists exhibit little tendency toward fraud and outright deception of others (ibid., 276–77). Disinterestedness is thus an important ethos for the integrity of science, but it is not equivalent to being value-free. Rather than argue that science is a value-free enterprise, even in its ideal form, Merton suggests that "the ethos of science is that affectively toned complex of values and norms which is held to be binding on the man of science" (ibid., 268–69). This echoes what the value-free ideal would become, a strict adherence to internal norms only, but Merton saw such adherence in more complex terms, pointing out that any ethos of science needed to be in sync with the broader social context, and that scientists had to take responsibility for the social implications of their work, or risk undermining support for scientific research (Merton 1938, 263). It was a rejection of this aspect of Merton's argument that would solidify the modern value-free ideal.

One also cannot find clear or strong endorsements for value-free science among the developing philosophy of science community. At the start of World War II, the predominant philosophical view of science and values was that the two were inextricably intermixed in practice, and that no dichotomy between the two existed. John Dewey, America's most prominent philosopher and leader of pragmatism, viewed science as an essentially value-laden enterprise, dependent upon human needs and concerns for its direction and understanding. The logical empiricists, the European philosophers fleeing an increasingly hostile Central Europe, held similarly complex views on the subject. Although the English philosopher A. J. Ayer dismissed all values as obscurantist metaphysics with no place in a scientific understanding of the world, other prominent logical empiricists disagreed.[7] Many of them emigrated to the United States in the 1930s. Consider, for example, Rudolf Carnap, a central leader of the logical empiricists, who came to the United States in 1937. Carnap argued that one could conceptually separate semantics, syntax, and pragmatics, but he did not think one could pursue a particular project in semantics or syntax (ways of understanding language) without some pragmatic commitments driving the direction of the investigation. Such pragmatics necessarily involved value commitments (Reisch 2005, 47–51). In other words, no *full* account of scientific knowledge could exclude values. Other leaders such as Otto Neurath and Philip Frank were deeply interested in the interrelationships of science, values, and society and did not see them as fundamentally separate domains.[8]

Yet in the 1940s, this understanding of science as embedded in a society, as having a clear and direct impact on society, and of this relationship being of philosophical interest, began to be undermined. Philosophers such as Carl Hempel and Hans Reichenbach pursued a program of logical analysis of scientific reasoning, of attempting to explicate in abstract form the logical relationships between scientific theory and empirical evidence. Reichenbach argued that, philosophically, which theories we are testing is uninteresting and belongs to the purely psychological/social context of discovery, whereas philosophers had their work centered in the context of justification: what evidence would justify accepting any given theory.[9] Hempel (along with Paul Oppenheim) similarly focused on issues such as what the logic of confirmation is (how much confirmation is provided by a piece of evidence) and what logically constitutes a scientific explanation (see Hempel 1965a). These works became the technical core of philosophy of science for decades to come. They aimed at a rational reconstruction of actual scientific practice, of trying to understand the logical relations that ultimately justified the acceptance of scientific knowledge. As such, they ignored not only actual internal scientific practice, but also the relations between science and the society in which it operates.

The implications of this kind of work for the value-free ideal can be seen in Reichenbach's *The Rise of Scientific Philosophy* (1951), where Reichenbach argues that knowledge and ethics are fundamentally distinct enterprises, and that confusions between the two had plagued the history of philosophy, leading much of it astray. In his efforts to keep knowledge and ethics distinct, he declares, "The modern analysis of knowledge makes a cognitive ethics impossible" (277). Cognitive ethics was the idea that ethics are a form of knowledge; thus, according to Reichenbach, ethics cannot be a form of knowledge. He continues, "knowledge does not include any normative parts and therefore does not lend itself to an interpretation of ethics" (277). In Reichenbach's view, knowledge is purely descriptive, ethics is purely normative, and the two have nothing in common. If values are also purely normative, then under this view knowledge has no values involved with it at all. Science, the epitome of knowledge for Reichenbach, would necessarily be value free.

Reichenbach's view did not go unchallenged, as we will see in the next section. But a focus on the logic of science, divorced from scientific practice and social realities, was an increasingly attractive approach for the philosophy of science as the cold war climate intensified. Although debate over whether and how science and values were related continued in the pages of the journal *Philosophy of Science* (see Howard 2003, 67), a progressive narrowing of interest in the field and an exclusion of complicating realities began. The broad range of ideas concerning science and values that held sway in the 1930s and early 1940s was no longer acceptable. The cold war demanded strict dichotomies: either one was with the United States, its capitalism, and its democracy, or one was with the Soviet Union, its communism, and its totalitarianism. Middle paths were unavailable. In this climate, arguing for links between science and values was uncomfortable for several reasons. First, in Marxist philosophy, science and values are closely interrelated. Arguing for any such links put one's work under the scrutiny of anticommunist crusades, whether or not one's work was Marxist.[10] Second, discussions concerning the relationship between science and society tended to be broadly based. There was no canonical literature on which arguments were built; there was no agreed starting point for discussion. The widespread response of academics to the pressures of anticommunist crusades, which were under way by 1948 and lasted into the 1960s, was to professionalize their fields, narrowing their expertise and focusing on a well-defined topic.[11] The topic of values in science made such a narrowing difficult. Yet the desire to professionalize the discipline of philosophy of science was palpable by the late 1950s.[12] Third, even if one could perform such a narrowing concerning science and values, it would be prudent and preferable if philosophy of science looked as apolitical as possible. For philosophers of science, creating a professional domain where science was essentially or ideally apolitical would help philosophers survive the cold war unscathed. Setting up an ideal for science as value free was central to creating an apolitical disciplinary arena for philosophy of science. Yet the value-free ideal was not obviously acceptable in the early 1950s.

The 1950s Debate on Values in Science

Even as the cold war deepened, a debate, the last robust debate, about values in science developed. In the early 1950s, Reichenbach's call for a scientific (and thus value-free) philosophy was not uncritically accepted. In *The Philosophical Review*, Norman Malcolm lambasted Reichenbach's *The Rise of Scientific Philosophy*, arguing that Reichenbach "exaggerates the importance of symbolic logic for philosophy," and that Reichenbach's own arguments show "that 'the logician' and 'scientific philosopher,' just as anyone else, can fall into primitive nonsense" (Malcolm 1951, 585–86). Even the supportive review by R. F. J. Withers (1952) in the *British Journal for the Philosophy of Science* found the book unconvincing in the end. So although the views of philosophers such as Reichenbach and Ayer were widely discussed, and a narrow technical focus for philosophy of science was gaining support, the value-free ideal was not yet in place or fully developed.[13] In reaction to the growing tide of technical, narrowly construed philosophy of science, some philosophers argued for the need for values in science. The responses to their ideas helped to generate the value-free ideal that held sway after 1960.

The most prominent arguments against the view of science as a value-free enterprise were developed by C. West Churchman, Philip Frank, and Richard Rudner. Between 1948 and 1954, both Churchman and Rudner presented widely read arguments that social and ethical values are required components of scientific reasoning, while Frank supported these endeavors. In their view, the scientist as public advisor and decisionmaker was an important role for scientists, and this role necessitated the use of ethical values in scientific reasoning. The Churchman-Rudner arguments forced philosophers of science into a dilemma: either accept the importance of values in science or reject the role of scientist as public decisionmaker. Rather than accept values in science, the field chose the latter option, thus rejecting a careful consideration of the public role of science. This allowed the value-free ideal to take up its position as the widely accepted doctrine it became: that scientists should consider only internal values when doing science.

Pragmatics of Induction: The Last Gasp for Values in Science

In general, the arguments of Churchman and Rudner centered on the issue of how much evidence a scientist should require before accepting (or rejecting) a particular scientific claim or hypothesis. Both Churchman and Rudner suggested that the amount of evidence can and should change, depending on the context in which the accepting or rejecting took place. Thus, depending on the consequences of a wrong choice (which is contingent on the context of choice), scientists could demand more or less evidence before coming to accept or reject a view. The weighing of the consequences required the use of values in the scientist's choice.

This emphasis on context was quite different than the approach of some philosophers (for example, Hempel and Oppenheim) who focused primarily on the nature of confirmation in science in the late 1940s. Confirmation studies dealt with how much a piece of evidence lent support to a particular scientific hypothesis. While Churchman was interested in issues of confirmation, he found such questions insufficient for understanding all of scientific practice, and thus began to ask under what conditions a scientist ought to *accept* a hypothesis, not just consider it *confirmed*. While the degree of confirmation was still important, Churchman (1948a) argued that it was insufficient for deciding whether to accept or reject a hypothesis (see Churchman 1948a and 1948b): "But there would be cases where we would not want to accept an hypothesis even though the evidence gives a high d.c. [degree of confirmation] score, because we are fearful of the consequences of a wrong decision" (Churchman 1948a, 256).

Churchman argued that one must consider the ends to which one will use a hypothesis in order to make a full evaluation of the hypothesis. Because there are a range of ends to which one can use scientific hypotheses, including uses outside of science proper, Churchman believed that "the complete analysis of the methods of scientific inference shows that the theory of inference in science demands the use of ethical judgments" (265). We must have "ethical criteria of adequacy" in order to find that some theory or hypothesis is adequately supported.

An essay by Richard Rudner published five years later in *Philosophy of Science* developed Churchman's argument further. Rudner (1953) argued that social and ethical values were often essential to complete scientific reasoning. He argued for this by making two uncontroversial claims: (1) "that the scientist as scientist accepts or rejects hypotheses," and (2) that "no scientific hypothesis is ever completely verified" (2). With these two claims in hand, Rudner stated,

> In accepting a hypothesis the scientist must make the decision that the evidence is sufficiently strong or that the probability is sufficiently high to warrant the acceptance of the hypothesis. Obviously our decision regarding the evidence and respecting how strong is "strong enough," is going to be a function of the *importance*, in the typically ethical sense, of making a mistake in accepting or rejecting the hypothesis. . . . *How sure we need to be before we accept a hypothesis will depend on how serious a mistake would be.* (2)

According to Rudner, determining the seriousness of a mistake requires a value judgment, and thus value judgments play an essential role in science, in shaping the acceptance or rejection of a hypothesis.

Rudner saw his work as a direct challenge to the eschewal of serious philosophical examination of values advocated by some analytic philosophers:

> If the major point I have here undertaken to establish is correct, then clearly we are confronted with a first order crisis in science and methodology. . . . What seems called for . . . is nothing less than a radical reworking of the ideal of scientific objectivity. . . . Objectivity for science lies at least

> in becoming precise about what value judgments are being and might have been made in a given inquiry—and even, to put it in its most challenging form, what value decisions ought to be made; in short . . . a science of ethics is a necessary requirement if science's progress toward objectivity is to be continuous. (ibid., 6)

Because of the implications of accepting a scientific hypothesis, science is drawn inexorably into ethics, and vice versa. Certainly science should not be value free, according to Rudner. While I do not agree with Rudner that we need a science of ethics in order to procure objective science, Rudner clearly saw that his arguments were a challenge to the growing antipragmatism in philosophy of science, as analytic philosophy rose in prominence. If philosophers and scientists were to take seriously the practical implications of inductive arguments, a conceptualization of objectivity that allowed for values would be needed. Minimally, the values used to make decisions would have to become an explicit part of scientific discussion.

The views of Churchman and Rudner were advanced in several settings and taken quite seriously at the time. At the December 1953 meeting of the American Association for the Advancement of Science (AAAS), a special set of sessions on the "Validation of Scientific Theories" was arranged by the Philosophy of Science Association, the Institute for the Unity of Science, and the History and Philosophy of Science section of AAAS, and sponsored by the newly formed National Science Foundation. The first session, on the "Acceptance of Scientific Theories," focused primarily on the work of Churchman and Rudner, with Churchman and Rudner presenting their main arguments regarding the necessity of values in scientific reasoning. (The papers were reprinted in *Scientific Monthly*, September 1954.) In his opening talk for the session, Philip Frank set the stage for Churchman and Rudner by acknowledging the complex set of goals for science:

> The conviction that science is independent of all moral and political influences arises when we regard science either as a collective of facts or as a picture of objective reality. But today, everyone who has attentively studied the logic of science will know that science actually is an instrument that serves the purpose of connecting present events with future events and deliberately utilizes this knowledge to shape future physical events as they are desired. (Frank 1953, 21–22)

In other words, science is not solely about truth or facts but also about shaping our world in a particular way. And in his essay based on this talk, Frank suggests that there are multiple possible ends that science can serve, multiple ways in which we might want to shape our world. Not only are values a part of science, but science is very much tied to society and our decisions about its future.

Although there were no critiques of the Churchman-Rudner position at the AAAS session, it was not long before other philosophers responded to their arguments. These challenges led directly to the isolated view of science, generating the current ideal of value-free science.

Closing the Ranks: Removing Social Values from Science

The program for philosophy of science that Rudner and others proposed was opposed by those who wanted a narrower focus for philosophy of science. How to stymie the force of these arguments? Two different positions were taken, both of which bolstered the isolated view of science that would become standard for philosophy of science in the ensuing decades. Both lines of critique suggested that scientists should not think beyond their narrow disciplinary boundaries when deciding what to accept or reject as sufficiently warranted science. The first line of argument suggested that scientists do not accept or reject hypotheses at all, but merely assign probabilities. If tenable, this argument would keep scientists from having to think about the possible consequences of error. However, this position had serious flaws that made it unsatisfactory for most. The second line of argument became widely adopted. With this position, scientists are not to think beyond their own scientific communities when evaluating their work. Philosophers accepted the idea that scientists must use values when making decisions in science, but rejected the notion that these values should include broad social or ethical values. Instead, the only legitimate values were narrow, disciplinary ones. By isolating scientists from social concerns and from the social implications of their work, philosophers were able to define and defend the ideal of value-free science.

The first line of argument, that scientists neither accept nor reject hypotheses, was developed by Richard Jeffrey.[14] His 1956 "Valuation and Acceptance of Scientific Hypotheses" argues that scientists should neither accept nor reject hypotheses. Instead, scientists should assign probabilities to hypotheses, and then turn the hypotheses and their assigned likelihoods over to the public. However, Rudner (1953) had already considered this position, arguing that even if this were the case, even if scientists only assigned probabilities to hypotheses and left the outright acceptance or rejection to their audience, the scientist would have to reject or accept the probability attached to the hypothesis, thus simply pushing the problem back one step (4). The scientist must say something, whether outright acceptance/rejection or a probability judgment. Either way, the scientist must decide that there is sufficient warrant for their stand, a decision that requires consideration of values in order to weigh the potential consequences of error. Jeffrey (1956) acknowledges Rudner's argument on this point: "Rudner's objection must be included as one of the weightiest [for the] . . . probabilistic view of science" (246). Despite his acknowledgment of the problem, Jeffrey provided no direct response to Rudner's argument.[15]

Although few have followed Jeffrey in arguing that scientists should not accept or reject hypotheses, several of Jeffrey's arguments were developed to pursue the second line of argument, that scientists should not consider the consequences of error beyond the narrow confines of the scientific community. First, Jeffrey suggests that one could never tell what the seriousness of a mistake in the acceptance or rejection of hypotheses might be. He argues that in general the acceptance or rejection of scientific hypotheses involves such a complex mixture of potential consequences from mistakes that it is unreasonable to expect scientists to consider them all: "It is certainly meaningless to speak of *the* cost of mistaken acceptance or rejection, for by its nature a putative scientific law will be relevant in a great diversity of choice situations among which the cost of a mistake will vary greatly" (Jeffrey 1956, 242). Thus, it is useless to demand that scientists consider these costs. Jeffrey even argues that the acceptance or rejection of hypotheses (and the consideration of values involved) can even be harmful: "If the scientist is to maximize good he should refrain from accepting or rejecting hypotheses, since he cannot possibly do so in such a way as to optimize every decision which may be made on the basis of those hypotheses" (ibid., 245).

Scientific pronouncements can be used in so many contexts, Jeffrey suggests, that the values used to accept a hypothesis in one context could be quite damaging when applied to another. Jeffrey provides no discussion of alternative approaches, such as making the value judgments explicit so that scientific work can be properly applied, or directly involving scientists in particular decisionmaking contexts where their work is used. Jeffrey's response to this complexity was to sever scientists from the context in which their hypotheses are used, and to remove them from considering the potential consequences of error altogether.

This idea, that scientists were too removed (or should be so removed) from the practical effects of their choices for social values to legitimately play a role in their decisionmaking, was developed further in the second line of argument against Rudner and Churchman, and was eventually codified in the value-free ideal. The crucial idea was that one could accept that values are needed for judgments concerning the acceptance or rejection of hypotheses, but limit the scope of those values. Churchman, in his 1956 defense of Rudner's essay from Jeffrey's critique, inadvertently opens this avenue of debate when he suggests that many of the decisions made by scientists on whether to accept or reject hypotheses could be justified in terms of values internal to the scientific process and its aims. Scientists whose work has no clear, practical implications would want to make their decisions considering such things as: "the relative worth of (1) more observations, (2) greater scope of his conceptual model, (3) simplicity, (4) precision of language, (5) accuracy of the probability assignment" (Churchman 1956, 248).

But Churchman did not think such "epistemic values" (as philosophers came to call them) were necessarily the *only* considerations for *all* scientists. This was precisely the position, however, presented by Isaac Levi. Levi argued that scientists should utilize only "epistemic values" in their judgments of whether there is sufficient evidence for accepting a hypothesis. According to Levi, scientists, by being scientists, submit themselves to scientific "canons of inference" which limit the scope of values for consideration:[16]

> When a scientist commits himself to certain "'scientific" standards of inference, he does, in a sense, commit himself to certain normative principles. He is obligated to accept the validity of certain types of inference and to deny the validity of others. . . . In other words, the canons of inference might require of each scientist *qua* scientist that he have the same attitudes, assign the same utilities, or take each mistake with the same degree of seriousness as every other scientist. . . . [It is not the case] that the scientist *qua* scientist makes no value judgment but that given his commitment to the canons of inference he need make no further value judgments in order to decide which hypotheses to accept and which to reject. (Levi 1960, 356)

The canons of inference limit the values scientists can and should use in evaluating whether to accept or reject hypotheses. If the sole goal of science is "to replace doubt by true belief," then "epistemic values" (such as simplicity, explanatory power, and scope) are sufficient for setting decision criteria for scientists, and scientists should not go beyond those values (Levi 1962, 49). Levi was deliberately severing the scientists from the broad social context of decisionmaking that concerned Rudner. Scientists, when using only the canons of inference, should not think about the potential social implications of their work or of potential errors, or consider social or ethical values in the acceptance or rejection of scientific theories. I will argue in chapter 5 that in many of the examples from science relevant to public policy, internal epistemic values are not helpful in weighing competing hypotheses. Nevertheless, in Levi's view, scientists need not and *should not* think beyond their own disciplinary boundaries when making judgments. The Rudner-Churchman arguments were rejected in favor of this value-free ideal. Levi's value-free ideal, that only epistemic values should be considered when doing scientific research, became the standard position for the next several decades.[17]

Embracing the Value-Free Ideal

Levi articulated a way in which philosophers of science could defend a value-free ideal for science. If philosophers of science adopted the ideal, they could more carefully demarcate the boundaries of the newly forming discipline of philosophy of science while insulating their nascent discipline from political pressures.[18] Yet philosophers of science did not immediately embrace the value-free ideal. In the early 1960s, two prominent philosophers, Ernst Nagel and Carl Hempel, discussed the role of values in science and seemed to acknowledge the import of the Rudner-Churchman argument. Their views on the role of values in science expressed some ambivalence toward Levi's value-free ideal. The ultimate acceptance of the value-free ideal rested more on the shifting attention of the young discipline than on a decisive conclusion to the 1950s debate.

Ambivalence over the Value-Free Ideal

In 1961, Ernst Nagel published *The Structure of Science: Problems in the Logic of Scientific Explanation*. As is evident by the title, the book fell squarely into the developing tradition of a more narrowly focused philosophy of science, which was concerned with the logic of explanation, the character of scientific laws, and the status of scientific theories. Toward the end of the book, Nagel addresses concerns arising from the social sciences. One of those concerns centers on the nature of value-based bias in social science, and whether it is an insurmountable problem for the social sciences (Nagel 1961, 485–502). Nagel also compares the issue of values in the social sciences with the issue of values in the natural sciences. In the end, he argues that none of the problems arising in the social sciences are truly peculiar to the social sciences.[19] Thus, his discussion reveals what he thought about values in science generally.

Nagel (1961) divided the potential places for the influence of values on science into four categories: "(1) the selection of problems, (2) the determination of the contents of conclusions, (3) the identification of fact, and (4) the assessment of evidence" (485). For all four possible roles, Nagel argues both that values often do play a role in science and that the issues for social science are no different than for natural science. The first role, values influencing the selection of problems, was a readily acknowledged and unproblematic place for values to influence science. Nagel notes that "the interests of the scientist determine what he selects for investigation," and suggests that "this fact, by itself, represents no obstacle to the successful pursuit of objectively controlled inquiry in any branch of study" (486–87). More problematic would be the second possible role, which Nagel describes as an inevitable infiltration of one's values into one's inquiry, even without one's awareness. To address this potential problem, Nagel presciently suggests, "The difficulties generated by scientific inquiry by unconscious bias and tacit value orientations are rarely overcome by devout resolutions to eliminate bias. They are usually overcome, often only gradually, through the self-corrective mechanisms of science as a social enterprise" (489).

In the 1990s and 2000s, philosophers such as Helen Longino focused on developing a philosophical account of these self-corrective social mechanisms.[20] At the time, Nagel found this issue to be equally problematic for both natural and social sciences. Similarly, for the third possible role, which Nagel explicated as whether values could be kept distinct from facts at all, he argued that this is a problem for both natural and social sciences and needs constant attention. Thus, neither of these two roles are presented as legitimate or laudable avenues for values in science (as the first role is), but as issues for vigilance.

The fourth role for values in science is central to our concerns here. Nagel (1961, 496–97) replayed the Rudner-Churchman arguments about the need for values in the assessment of the adequacy of evidence (although without citing Rudner or Churchman). After discussing the example of which kinds of errors should be more assiduously avoided in the testing of a promising drug for toxicity, Nagel noted that the appraisal of statistical testing could provide one with no fixed rule on such a question because it depends on the potential consequences of error in the particular context: "The main point to be noted in this analysis is that the rule presupposes certain appraising judgments of value. In short, if this result is generalized, statistical theory appears to support the thesis that value commitments enter decisively into the rules for assessing evidence for statistical hypotheses"

(Nagel 1961, 497). Nagel, however, is unsure how far such a result should be generalized. He continues: "However, the theoretical analysis upon which this thesis rests does not entail the conclusion that the rules actually employed in every social inquiry for assessing evidence necessarily involve some *special* commitments, i.e., commitments such as those mentioned in the above example, as distinct from those generally implicit in science as an enterprise aiming to achieve reliable knowledge" (497).

Here we have echoes of Levi's view, that scientists in many cases need not consider values beyond those that make up the "canons of inference" in scientific research, that is, epistemic values. But Nagel does not argue that scientists should only consider such epistemic values. Rather, he suggests that scientists often need not go beyond such considerations, not that they should not. This argument leaves open to the judgment of the scientist in a particular context which values are appropriate for the assessment of evidence. In the end, Nagel provided no strong endorsement of the value-free ideal in science. The scientist is not yet fully severed from the social context.

A similar ambivalence can be found in Carl Hempel's essay, "Science and Human Values" (1965b), which addresses all aspects of the relationship between science and values (first published in 1960, reprinted in 1965).[21] Much of the essay is devoted to the influence science can have on the values that we hold and the decisions we make.[22] Near the end of the essay, Hempel turns to the issue of what influence values should have on science, or "whether scientific knowledge and method presuppose valuation" (90). Hempel first rapidly narrows the scope of the question; he accepts the role of values in both the selection of questions to pursue and the choice of the scientist's career. He denies that values themselves can confirm scientific knowledge. However, the question of whether values play a role in scientific method, or in the doing of science, is a more complex issue.

In questioning whether values have a logical role to play in the doing of science, Hempel answers in the affirmative. In line with Rudner's argument, Hempel recognizes the need for values in the acceptance or rejection of hypotheses.[23] The need for values in science arises from the "inductive risk" inherent in science: "Such acceptance [of a scientific law] carries with it the 'inductive risk' that the presumptive law may not hold in full generality, and that future evidence may lead scientists to modify or abandon it" (Hempel 1965a, 92). Because the chance of error is always present, decision rules for the acceptance of scientific hypotheses are needed. And "the formulation of 'adequate' decision rules requires . . . specification of valuations that can then serve as standards of adequacy" (92). Adequate decision rules would consider the possible outcomes of accepting or rejecting a hypothesis, and these outcomes include the following four main possibilities: (1) the hypothesis is accepted as true and is, (2) the hypothesis is rejected as false and is, (3) the hypothesis is accepted as true but really is false, and (4) the hypothesis is rejected as false but really is true (ibid., 92). The problem of inductive risk lies in the latter two possibilities, where values must be used to decide the seriousness of error in those cases.

The issue then becomes what kind of values are needed to set up adequate rules of acceptance. For Hempel in the 1960s, the values depended on the case. For some situations, nonepistemic values would be required, particularly for the use of science in industrial quality control: "In the cases where the hypothesis under test, if accepted, is to be made the basis of a specific course of action, the possible outcomes may lead to success or failure of the intended practical application; in these cases, the values and disvalues at stake may well be expressible in terms of monetary gains and losses" (Hempel 1965a, 92–93).

Thus, under Hempel's model, social and ethical values, translated into a monetary or utility metric, can and sometimes do play an important role in scientific judgments. However, aside from such cases with clear applications, Hempel suggested that social and ethical values need not be considered. He saw most of scientific practice as being removed from practical implications, that is, as being "pure scientific research, where no practical applications are contemplated" (ibid., 93). Because of this, he argued that epistemic values usually suffice when making judgments about inductive risk in science. Yet social and ethical values do have a rational place in scientific reasoning for some areas of science, particularly those with clear practical import. In sum, Hempel and Nagel appear to have agreed on this view for the role of values in some science in the early 1960s.

In his later writings, Hempel occasionally struggled with the role of values in science. For example, in his "Turns in the Evolution of the Problem of Induction" from 1981, Hempel discusses the Rudner paper and Jeffrey's response explicitly (reprinted in Hempel 2001; see esp. 347–52). He notes that Jeffrey's response to Rudner rested on the view that scientists do not play a role in practical decisionmaking: "But, Jeffrey notes, the scientist *qua* scientist is not concerned with giving advice, or making decisions, on contemplated courses of action" (Hempel 2001, 348). However, Hempel ultimately disagreed with Jeffrey on this point (350). Despite apparently siding with Rudner in theory, in practice Hempel was primarily concerned with science with no practical import, with science solely concerned with "the increase of scientific knowledge" (350–51). Thus, he focused on the nature of epistemic values, trying to nail down the guidance they give (see also Hempel 2001, 372–95). In the practice of his work, Hempel had accepted the value-free ideal for science. But how did the philosophy of science community come to this tacit, and later explicit, endorsement of the value-free ideal from its apparent ambivalence in the early 1960s?

The Acceptance of the Value-Free Ideal

The shift from the acknowledgment of the need for social and ethical values in at least some scientific work, particularly work to be used for decision-making outside of science, to the value-free ideal for all of science required a crucial step: the viewing of the scientific community as demarcated and isolated from the surrounding society. Shortly after the first publication of Hempel's and Nagel's views on values in science, Thomas Kuhn's *The Structure of Scientific Revolutions* was published, the work that famously introduced the concept of scientific paradigms.[24] Although Kuhn historicized science in an exciting way, he also argued for the view that the scientific community was best understood as distinct and isolated from the surrounding society. In Kuhn's eyes such isolation was laudatory, and perhaps essential for the genuine character of science. When discussing how paradigms function in science, Kuhn (1962) notes that a "paradigm can . . . insulate the [scientific] community from those socially important problems not reducible to puzzle form" (37). A focus on solving puzzles—problems assumed to have a well-defined answer—is key to conducting normal science. Thus, insulation from socially important problems is crucial, in Kuhn's view, for the development of science. Paradigms, those defining frameworks for science, are to be local to particular sciences or disciplines, or at most to science itself, rather than representative of broad cultural trends. Paradigm shifts are driven by struggles internal to science. The demarcation between science and society is essential for science to be proper science.

In the closing pages of the book, Kuhn reemphasizes the importance of a science isolated from society. After discussing the nature of progress in science, Kuhn notes:

> Some of these [aspects of progress] are consequences of the unparalleled insulation of mature scientific communities from the demands of the laity and of everyday life. That insulation has never been complete—we are now discussing matters of degree. Nevertheless, there are no other professional

> communities in which individual creative work is so exclusively addressed to and evaluated by other members of the profession. . . . Just because he is working only for an audience of colleagues, an audience that shares his own values and beliefs, the scientists can take a single set of standards for granted. . . . Even more important, the insulation of the scientific community from society permits the individual scientist to concentrate his attention upon problems that he has good reason to believe he will be able to solve. (164)

Kuhn goes on to argue that because of this insulation from the needs and values of society, natural (not social) scientists are more likely to be successful in solving their puzzles and problems, and thus insulation from society is key to the success of natural science. With such a demarcation, Kuhn's discussion of the workings of science considers the scientific community to be a self-contained community whose problems and debates are discussed and decided *within* that community.

It is hard to underestimate the impact of Kuhn's book on the philosophy of science. Many philosophers developed alternative views of scientific change, the history of science, and scientific progress to counter Kuhn's account. The issue of scientific realism picked up steam in part in reaction to Kuhn's seemingly antirealist view of the history of science. The need for more robust historical accounts in the philosophy of science became a central aspect of the field, leading to programs and departments in the history *and* philosophy of science. Kuhn challenged the prevailing views in philosophy of science in many ways, and his work shaped the field for years to come. As philosophers grappled with some aspects of Kuhn's work, others were left unchallenged, including Kuhn's isolated view of science, which may have been widely accepted in part because philosophers were already leaning toward that idea. The field of philosophy of science was greatly simplified and streamlined if science was thought of as cut off from the society in which it operated.

For the philosophers who wrote in the wake of Kuhn 1962, attention focused on problems conceptualized as wholly internal to science, such as the nature of scientific change, the realism debate, or problems internal to special areas of science. The issue of values in science came to be about the nature of epistemic values, and only epistemic values, in science. This can be seen in the later work of Hempel and the work of Larry Laudan, whose *Science and Values* (1984) is solely concerned with epistemic values.[25] Laudan acknowledges Kuhn's influence in this regard, noting that Kuhn was "ever a believer in a sharp demarcation between science and everything else" (xiii). This aspect of Kuhn was imitated by many of his readers, even as they criticized him for not having an adequately rational account of scientific change, or a rigorous enough definition of a paradigm. And it was this image of science, as demarcated from everything else, that was the final piece needed to cement the value-free ideal. For with Kuhn's isolated view of science in place, Levi's argument (1960) that scientists should not think beyond the canons of scientific inference took on the veneer of the obvious—of course scientists should only think about issues within the realm of science.

Not all philosophers of science accepted Kuhn's understanding of science. Challenges to an isolated view, through challenges to the value-free ideal, did occur occasionally.[26] Leach 1968 and Scriven 1974 are two notable examples.[27] But the arguments in these papers were largely ignored. A similar fate befell James Gaa's "Moral Autonomy and the Rationality of Science" (1977), which critiqued an isolated understanding of science. Gaa notes the importance of presupposing that science is isolated from the rest of society for the value-free ideal and describes moral autonomy as the view that "acceptance (and rejection) decisions ought to be made with regard only to the attainment of the characteristic ('purely epistemic') goals of science" (514). Despite its importance for philosophy of science and the value-free ideal, Gaa notes that "the moral autonomy of science is a largely unexamined dogma" (514). In his examination of this dogma, Gaa suggests that it is not obviously sound: "Even if science does have characteristic goals—goals to distinguish it from other kinds of activities—scientists ought, in their acceptance decisions, to consider more than the utility of a theory for those special goals" (525). In the end, Gaa found the arguments for such a moral autonomy (as much as there were any) unconvincing, particularly in the face of one possible conclusion resulting from moral autonomy, that "scientists can be both *rational* and *irresponsible (immoral)* at the same time" (536). This would occur because scientists would only consider Levi's internal canons of inference, and not the broader implications of their decisions for the society in which science functions. Such a view of scientific rationality was not acceptable to Gaa. Unfortunately, Gaa's detailed examination of this "dogma" went largely unnoticed and undiscussed.[28]

Despite Leach 1968, Scriven 1974, and Gaa 1977, and the rising importance of scientists in public life, philosophers of science continued to hold to the value-free ideal and the isolated view of science on which it rested. In 1982, when policymakers and scientists were struggling to find new ways to better incorporate science advice into the policy process, Ernan McMullin (1983) argued for the value-free ideal, rejecting Rudner's argument and echoing Jeffrey (1956) and Levi (1962):

> If theory is being applied to practical ends, and the theoretical alternatives carry with them outcomes of different value to the agents concerned, we have the typical decision-theoretic grid involving not only likelihood estimates but also "utilities" of one sort or another. Such utilities are irrelevant to theoretical science proper and the scientist is not called upon to make value-judgments in their regard as part of his scientific work. . . . The conclusion that Rudner draws from his analysis of hypothesis-acceptance is that "a science of ethics is a necessary requirement if science's happy progress toward objectivity is to be continuous." But scientists are (happily!) not called upon to "accept" hypotheses in the sense he is presupposing, and so his conclusion does not go through. (McMullin 1983, 8)[29]

Both the apparent obviousness of the scientist's removal from public decisionmaking and the sense of relief inherent in McMullin's comments are striking, particularly given the prevalence of scientists being called upon to give advice or to make decisions regarding which hypotheses are adequately supported in real and crucial contexts. Kuhn's isolated science was well entrenched in mainstream philosophy of science.

Even today, the value-free ideal continues to be the predominant view in the philosophy of science.[30] In Hugh Lacey's *Is Science Value-Free?* (1999), he argues strongly for the value-free ideal that he sees as being at the heart of science.[31] In order to defend the ideal from Rudner's arguments, he defines acceptance of a theory solely in terms of whether or not more research needs to be done on it, ignoring the multiple pragmatic aspects of acceptance, such as whether to use a theory as a basis for decisionmaking (Lacey 1999, 13–14).[32] Lacey sees the ideal as essential to science and raises substantial fears about what would happen if the ideal were rejected. Without the value-free ideal, Lacey warns, we lose "all prospects of gaining significant knowledge" (216). We would be thrown back on merely "wishing" the world to be a particular way or "the back and forth play of biases, with only power to settle the matter" (214–15). As I will argue in chapters 5 and 6, we can reject the value-free ideal without these dire consequences. However, Lacey is correct to worry about science being held under the sway of biases, or allowing wishing to make it so. We need another ideal and a more robust understanding of objectivity to go with it.

In sum, philosophy of science as a discipline has been profoundly shaped by the value-free ideal and still largely adheres to it. The depth of this commitment can be found in the lack of interest in serious critiques of the ideal, in the topics on which the field focuses its energies, and in the relief felt when the ideal can be left intact. Yet the ideal has not received sufficient scrutiny, particularly concerning whether it is an acceptable ideal. Gaa's concerns over the possibility for a rational but immoral science have not been addressed.

While these disputes might seem to be historically interesting only to philosophers of science, they also influenced the way in which scientists and policymakers understood the proper norms for science. This ideal tacitly or explicitly underlies the scientists' approach to their work. Concerns over bias, the role of advocacy in science, and the possibility of junk science all attest to the importance of the ideal for scientists as well as philosophers of science. For example, as noted in chapter 1, Kuhn's isolationist view of science plays a key conceptual role in Peter Huber's understanding of what constitutes a scientific fact, which then distinguishes sound science from junk science in Huber's account. Huber relies on Kuhn's view of the scientific community, concerning itself solely with epistemic values, to determine what is sufficiently reliable to be counted as a fact (Huber 1991, 226); only then does the isolated scientific community turn the "fact" over to society. However, this is not how science advising is structured, nor would it be desirable to have it so structured. Or consider how biologist Lewis Wolpert, in his *Unnatural Nature of Science*, reflects Levi's value-free ideal when he writes that "scientific theories may be judged in terms of their scope, parsimony—the fewer assumptions and laws the better—clarity, logical consistency, precision, testability, empirical support and fruitfulness," thus emphasizing the importance of epistemic values in science within the canons of scientific reasoning (Wolpert 1992, 17–18). Chemist Roald Hoffmann's "Why Buy That Theory?" (2003) also examines the list of classic epistemic values (simplicity, explanatory power, scope, and fruitfulness) as reasons to accept a theory, arguing that it is part of scientists' psychology to prefer theories with these attributes. His account reflects the classic Kuhn-Levi view that these values help scientists decide whether to accept a new theory in the arena of pure science. Admittedly, these pieces of evidence are at best suggestive; scientists rarely discuss the value-free ideal for science with the kind of precision found among philosophers of science, and so the influence of philosophy of science on these ideas is difficult to track. However, we do know that the value-free ideal appears to have influence beyond the philosophy of science.

Conclusion

The value-free ideal was not an obviously acceptable ideal at the beginning of the 1950s. Yet, even as Rudner and Churchman presented clear arguments for the importance of values in science, pressures to professionalize the young discipline of philosophy of science took hold. Arguments that scientists should only consider values internal to science were made. The idealized image of the isolated scientific community gained prominence, reinforced by Kuhn 1962. Although some criticisms of the value-free ideal have persisted, they have been ignored or marginalized by the philosophers of science. And the ideal has been influential among scientists as well as among philosophers of science.

The great irony of this history is that the ideal rests on a faulty presupposition, that science is isolated (or should be isolated) from the rest of society. The account in chapter 2 of scientists gaining ever more importance in the policy process belies the descriptive veracity of this presupposition. But despite the increased importance of science for society, one still might wonder whether science should aim to be value free. In other words, despite the deep ties and interrelationships between science and society, should scientists attempt to follow the value-free ideal, excluding consideration of the consequences of error in their work? Should they rely solely on the canons of inference internal to science? Should science be morally autonomous in this way? These are fundamentally questions concerning the moral responsibilities of scientists. It is to these questions that we will now turn.

CHAPTER 4

THE MORAL RESPONSIBILITIES OF SCIENTISTS

THE DEBATE AMONG PHILOSOPHERS OF SCIENCE in the 1950s concerning values in science hinged on the proper role of scientists in a modern democracy. Should scientists be giving advice to decisionmakers? And should they, when giving this advice, consider the context of use and the potential consequences of error when deciding what to say? Or should scientists decide which empirical claims are adequately supported with no thought to the importance of these claims to society? These questions fundamentally concern the moral responsibilities of scientists as scientists. If, with Rudner and Churchman, one thinks that scientists should consider the potential consequences of error when deciding which claims to make, then values have an unavoidable place in scientific reasoning. If, with Levi and McMullin, one thinks that scientists should not be considering the potential consequences of error, then scientists can safely exclude social and ethical values from the heart of scientific reasoning.[1]

Which is the correct view? Should scientists consider the potential consequences of error in their advising? This question involves two general aspects. First, there is the question of whether all of us, as general moral agents, have a responsibility to consider the consequences of error when deliberating over choices, and in particular when deciding upon which empirical claims to make. I will argue here that we do have a general moral responsibility to consider the consequences of error, based on our concern over reckless or negligent behavior. Second, there is the question of whether scientists share this burden with the rest of us, or whether they have a special moral exemption from such considerations. In other words, we must ask whether scientists have a special professional status which means they should not consider the consequences of their work as scientists. This is a view that has been supported by some in the recent past and must be addressed seriously. In the end, I will argue that scientists do have a moral responsibility to consider the consequences of error in their work, but that this responsibility places no burden of special foresight on the scientists. We cannot expect scientists to be seers. Indeed, the very errors we are concerned with here mean that we cannot expect *perfect* foresight and prediction. But we should expect *reasonable* foresight and care from our scientists. Being a scientist provides no special exemption from this expectation.

Moral Responsibility and the Consequences of Error

The literature on moral responsibility has developed much in recent years, but it has mostly focused on three general issues: competence (when is a person morally capable of making decisions and thus is responsible for them), coercion (what kind of forces on a person make their choices not their own, but rather due to someone else, thus shifting moral responsibility), and causation (what conception of causality allows for both enough free will and enough foresight so that we can be considered responsible for the outcomes of our actions).[2] While the discussion around these issues is fascinating, none of it is very illuminating for our concerns here. With respect to competence, scientists are generally capable moral agents, as capable in their daily lives as the rest of us. No one has seriously argued that scientific training somehow impairs moral reasoning or moral sentiments. With respect to coercion, scientists are not usually under threat of force to *not* consider certain moral issues. Coercion would be considered as pathological in science as it would anywhere else. And the issue of causation applies just as much for scientists as for anyone else; either we have a causal structure that allows for moral responsibility, or we do not. More relevant to our concerns here are what is moral responsibility in general and what are our responsibilities with respect to consequences we do not intend to cause. Let me address each of these in turn.

What do we mean when we say a person is morally responsible for some action or for an outcome of an action? One basic distinction is between causal responsibility and moral responsibility. I may be causally necessary, and thus partially causally responsible, for the eventual actions of my great-grandchildren, but few would suggest I was morally responsible for them, that I should be praised or blamed for what they do. Or if this book, for which I am causally responsible, serves to block a bullet meant for a reader of the book, I am surely not morally responsible for having saved that person's life, and no one should praise me (on that basis) for having written this particular book. Being part of a causal chain is not sufficient for moral responsibility. Where does moral responsibility begin? One crucial marker for moral responsibility is the giving of praise or blame. When we hold someone morally responsible for something, we want to give praise to them or lay blame on them. We think that they could have done otherwise, that they chose a particular course, and we hold them responsible for that choice.[3] Thus, the attribution of praise or blame is central to moral responsibility. In general there should be broad symmetries between what warrants praise and what warrants blame. In other words, if there are circumstances where we are willing to praise someone for some particular action or result, for those same general kinds of circumstances we should also attribute blame. The crucial point here is that the attribution of praise or blame is not equivalent to the attribution of cause. While some kind of causal connection is necessary, we are not held morally responsible for all the things in which we play a causal role.

When does a causal responsibility turn into a moral responsibility? Minimally, we are morally responsible for those things we intend to bring about. In these cases, a person chooses to do something deliberately, either because they think it is inherently the right (or wrong) thing to do or because of a particular sought consequence. The deliberate choice brings on the moral responsibility. Thus, if I intend to help or harm someone and I succeed, I am morally responsible in both cases, and usually praiseworthy in the former, blameworthy in the latter.

However, we are not morally responsible merely for those things we intended to bring about. We are also morally responsible to some extent for side effects of our actions. While this is widely accepted, it is a difficult question under which circumstances and to what extent we should be responsible for unintended consequences. Two general categories cover unintended consequences: recklessness and negligence. In discussing these terms, I follow Joel Feinberg's work and general legal usage. Feinberg (1970) writes, "When one knowingly creates an unreasonable risk to self or others, one is reckless; when one unknowingly but faultily creates such a risk, one is negligent" (193). When one is *reckless*, one is fully aware of the risks one is taking or imposing on others, and those risks are unjustified. What justifies certain risks can be contentious, particularly when the risks involve

other people not given decisionmaking authority or choice in a situa
Nevertheless, there are clear examples of justified risk (speeding on city streets to get a seriously injured person to the hospital) and unjustified risk (speeding on city streets for the fun of it). The key point is that we expect moral agents to carefully weigh such risks and to determine whether they are, in fact, justified.

If, on the other hand, one is not aware that one is risking harm, but one *should* be aware, then one is being *negligent*. When being negligent, one does not bother to evaluate obvious risks of harm, or one does not think about potential consequences of one's actions. As Feinberg notes, there are many ways in which to be negligent:

> One can consciously weigh the risk but misassess it, either because of hasty or otherwise insufficient scrutiny (rashness), or through willful blindness to the magnitude of the risk, or through the conscientious exercise of inherently bad judgment. Or one can unintentionally create an unreasonable risk by failing altogether to attend either to what one is doing (the manner of execution) or to the very possibility that harmful consequences might ensue. (Feinberg 1970, 193–94)

The difficulty with negligence, in addition to determining whether a risk is justified, is to determine what should be expected of the agent. How much foresight and careful deliberation should we expect the individual to have? Often, this question is answered through an examination of community standards, couched in terms of what a *reasonable person* would have done in like circumstances.

Through recklessness and negligence, one can be held morally responsible for unintended consequences both when things go the way one expects them and when things go awry. Through negligence, things may go exactly as planned (as far as you planned them), and still harmful and clearly foreseeable consequences would be your fault. You would be responsible because you should have foreseen the problems and planned further. For example, suppose you set fire to a field one dry summer to clear it of brush. You didn't bother to think about how to control the fire, not recognizing the obvious risk. Because of your negligence, harm caused by the fire raging beyond your property is your moral responsibility. If, on the other hand, you are aware of the risks in setting the fire, decide not to care and proceed anyway, then you are morally responsible for the damage when the fire escapes because of recklessness. The distinction between recklessness and negligence thus rests on the thought processes of the agent, on whether they reflect on potential consequences (either if events go as planned or unexpected "errors" occur), and on whether there was any attempt to prevent or mitigate possible harms that could arise from the chosen action. Recklessness is proceeding in the face of unreasonable risk; negligence is the failure to foresee and mitigate such risk.

This discussion of moral responsibility for choices might seem far removed from the context of science, where one is primarily concerned with coming to the correct empirical beliefs (and thus the province of theoretical instead of practical reason). However, science is a social enterprise, involving not just isolated individuals coming to have beliefs, but social communities working together to develop empirical work (Longino 1990, 2002; Solomon 2001). Even more pertinent, scientific work is developed and discussed within a society that takes the claims made in the name of science with special authority—hence the importance of science advising. Making empirical claims should be considered as a kind of action, with often identifiable consequences to be considered, and as a kind of belief formation process. We can consider everyday nonscientific examples to show that the responsibility to consider consequences of error embodied in concerns over recklessness and negligence applies not just to direct interventions in the world, but also to the making of claims, including descriptive claims.

For example, suppose one sees an unattended briefcase. Should one report the possible presence of a bomb? There are clear risks of error for making the descriptive claim that a bomb may be present. If one does not report it and it is a bomb, death and destruction may result. If one does report it and it is not, disruption of people's daily lives and distraction of resources away from more serious problems may result. Obviously, the former is a more serious consequence than the latter, but one should also weigh the uncertainties involved. Suppose, for example, that the briefcase is spotted in a busy subway station. In that context, where leaving a briefcase seems ominous, one should report the unattended briefcase. This is the reasonable weighing of risk and uncertainty in this context. Consider, however, a briefcase left in a college classroom, where the classroom is known to be used by a particularly absentminded colleague. In that context, while the consequences are similar, the uncertainties shift, and it is far more likely it is the colleague's briefcase than a bomb. Checking with the colleague first is the more prudent measure. In both cases, we expect each other to reflect upon the risks of making a claim, particularly the consequences of error and the uncertainties and likelihoods involved. Thus, we can be negligent or reckless in the making of descriptive or empirical claims.

This brief sketch of moral responsibility holds several crucial insights. Moral responsibility involves the attribution of praise or blame (and sometimes both). We are held morally responsible for our intentional choices and the intended consequences of those choices, but also for some (not all) of the unintended consequences. Those consequences that are reasonably foreseeable (more on this below), even if not foreseen by the individual, due to negligence, or even if ignored, due to recklessness, can be evaluated and the person held morally responsible for them. And finally, we are morally responsible for reflecting upon the possible negligence or recklessness of making claims, particularly when the making of a claim will have clear consequences for others. These are the standards we all live with in our daily lives, that help to shape our choices. The question is whether these standards should apply to scientists, and if so how, particularly in their role as advisors and public authorities.

Scientists and Moral Responsibility

What are the moral responsibilities of scientists? If scientists have the same responsibilities as the rest of us, they have the basic responsibilities we all share for the intended consequences of their choices, as well as for some of the unintended consequences. Specifically, they are responsible for the foreseeable consequences of their choices, whether intended or not. Thus, scientists have the responsibility to be neither reckless nor negligent in their choices. Because the impacts of science have such substantial, and often foreseeable, reach, meeting these basic responsibilities could profoundly influence the practice and direction of science.

We can set aside here the rare cases of scientists with genuinely malicious intent, and thus the problem of failing to meet the general moral responsibility of having a good aim, of one's intended consequences being morally acceptable. (Scientists can be as mistaken about the goodness of an aim as anyone else, but this poses no special problems for us here.) Even scientists who solely concern themselves with the pursuit of knowledge can be said to pursue a good, as it seems clear knowledge in general is thought to be a good. Unintended consequences are the more important— and murkier—ground here, even if we restrict our discussion to foreseeable consequences.

There are two kinds of unintended foreseeable consequences that may be of concern to the scientist. The first is the consequence that will likely result as a side effect even if the knowledge produced is perfectly reliable and accurate. Consequences to human subjects of research have been a focus of research ethics for over half a century, and they place restrictions on what kinds of research projects are morally acceptable. Even if the research goes exactly as planned, and the scientists intend the best, moral concerns may trump epistemic drives. This kind of concern takes on trickier overtones when one considers the impact of knowledge itself on society. Recent concerns have been raised over knowledge that, even if produced ethically and even if true, may harm society. This problem, what has been called the problem of forbidden knowledge, has received some attention lately, but discussions have stopped short of arguing that the potential harms to society are sufficient grounds for forbidding the pursuit of such knowledge (Johnson 1996, 1999; Kitcher 1997). Because these discussions center more on policy for science and democratic input into research agendas than on science for policy and its implications for scientific reasoning, this issue will be set aside here. More central to our purposes are the potential unintended consequences of making inaccurate or unreliable empirical claims, the second kind of unintended consequence. Given the public authority scientists wield, and the understanding of making empirical claims as a public action, there can be clear consequences for making a well-intended but still ultimately incorrect claim, just as with the briefcase example above. Thus, it seems that scientists can be reckless or negligent if they improperly consider the potential consequences of error based on their basic moral responsibilities.

These basic or general responsibilities are not the only ones relevant for scientists, however. There are also the responsibilities that scientists must meet because they are scientists. Responsibilities special to science can be considered *role responsibilities*, which are those that assist scientists in achieving the central goals of science: improved explanations and predictions about the world around us. They include the precepts of basic research ethics, such as the honest reporting of data, the open discussion of scientific results, and the fair consideration and evaluation of the work of others. There is little or no debate about whether these kinds of role responsibilities should hold for science. For example, all fraudulent reporting of data is roundly condemned. Debate centers, rather, on the extent of fraud in science (see, for example, Callahan 1994). Concern has also been raised about whether the rise of propriety research and the importance of business interests in science pose a threat to scientific integrity (Maker 1994). Few dispute that basic research ethics are central to the good functioning of science.

How should we understand the relationship between general or basic responsibilities and role responsibilities? Role responsibilities arise when we take on a particular role in society, and thus have additional obligations over and above the general responsibilities we all share. For example, when one becomes a parent, one takes on the additional responsibility of caring for one's child. This additional responsibility does not excuse one from the general responsibilities we all have, but must be met in conjunction with one's general responsibilities.[4] Thus, role responsibilities usually expand one's set of responsibilities. Examples range from family relationships to professional obligations. For scientists, the role responsibilities add the extra burden of care in reporting results and properly dealing with colleagues and students, for example. It is possible, but rare, that role responsibilities call for a contraction of general responsibilities. For example, consider a defense lawyer who has learned of past criminal activity from a client. The lawyer, unlike the rest of us, is under no obligation to report such activity because of the need to protect lawyer-client confidentiality. This exemption for the lawyer is possible only because of the clear structure of our criminal justice system. It is the responsibility of others to discover and investigate the past criminal activity. In the rigid adversarial system of our criminal justice system, the defense lawyer has the responsibility of being a client advocate, while others must discover and prosecute crimes, although even defense lawyers must report knowledge of crimes ongoing or not yet committed. Only within this kind of rigid system with clearly defined roles covering all the important responsibilities can a role responsibility lead to a reduction of general responsibilities.

Scientists do not currently work within such a rigid system with clearly defined roles covering all important responsibilities. Would scientists want to consider placing their work within such as system? In other words, would it be possible and desirable for someone other than a scientist to shoulder the burden of the general responsibility to consider the consequences of error in the scientist's work, so that the scientist does not have to worry about being negligent? There are several reasons such a burden-shifting is neither possible nor desirable. The most important reason is that it is doubtful anyone could fully take over this function for scientists. Because science's primary goal is to develop knowledge, scientists invariably find themselves in uncharted territory. While the science is being done, presumably only the scientist can *fully* appreciate the potential implications of the work, and, equally important, the potential errors and uncertainties in the work. And it is precisely these potential sources of error, and the consequences that could result from them, that someone must think about. The scientists are usually the most qualified to do so.

Despite the implausibility of scientists turning this aspect of general responsibility over to someone else completely, there is the possibility for some partial sharing of these general responsibilities. For example, in order to assure proper weighing of knowledge goals against the possible (even unintended) harms to humans in the pursuit of that knowledge, it is now standard practice for scientists using human subjects to submit their proposed methodologies to ethical review boards before proceeding with the experiment. Similar practices are becoming common for animal subjects as well. And I will suggest in chapter 8 that public involvement in some kinds of research can help to remove part of the burden from scientists as well. However, such sharing of the burden is the most that can be accomplished because scientists often encounter the unexpected and are the only ones aware of its presence, nature, and novelty. For scientists not to bear a general responsibility to consider potential unintended consequences of mistaken empirical judgment would require constant ethical oversight of all scientific practice. It is highly doubtful this could be accomplished (although moving toward such an option would surely mean full employment for applied ethicists), and it is even more doubtful scientists would prefer this option to shouldering most of the general responsibilities themselves. To abandon their general responsibilities would be to simultaneously relinquish most of their autonomy.

In order to see why, consider what would be required if we implemented a system that provided ethical oversight of all scientific decisions in order to remove the burden of these general responsibilities from scientists. The consideration of nonepistemic consequences could be neither an afterthought to the research project nor a process merely at the start of the project if the general responsibilities are to be properly fulfilled. Instead, such consideration would have to be an integral part of it and involved *throughout* the research project. Those shouldering the general responsibilities to consider social and ethical consequences of research (and in particular of errors in research) would have to have decisionmaking authority with the scientists in the same way that research review boards now have the authority to shape scientists' methodological approaches when they are dealing with human subjects. However, unlike these review boards, whose review takes place at one stage in the research project, those considering all of the nonepistemic consequences of scientific choices would have to be kept abreast of the research program at every stage (where choices are being made), and would have to have the authority to alter those choices when necessary. Otherwise, the responsibility would not be properly fulfilled and would not be able to keep pace with the developments accompanying discovery. Such intensive interference in scientific practice is anathema to most scientists.

In sum, there is no one currently available to shoulder the general moral responsibility to consider the consequences of error for scientists. Scientists themselves are the best qualified to do so, and to develop a body of people to take this burden from them would probably not be acceptable to scientists. This last point could use some emphasis. A cadre of ethical overseers would certainly draw the resentment of scientists. In order to thwart such oversight, scientists would likely become less thoughtful about both the presence of uncertainty in their work and the potential consequences of error. After all, if they do not identify the possibility of error, then there would be less need for oversight. Thus, not only would an oversight system be undesirable, it would likely be self-defeating.

On the face of things, it seems that scientists should meet both their role responsibilities and their general responsibilities. Despite the argument that no one can effectively take over the general responsibility to consider the consequences of error for scientists, some have suggested that scientists should nevertheless ignore or abandon such responsibilities. So, knowing that no one else can do the job, should scientists be shielded from the general responsibility to consider the consequences of error in their choices? When deciding how to proceed in a research project or considering which empirical claims to make, should scientists meet the general responsibilities discussed above? Some have thought that they should not.[5] As noted in chapter 3, many philosophers of science have assumed that scientists should be insulated (or act as if they were insulated) from such considerations. Others have made the claim more explicit. For example, Lübbe (1986) suggests that scientists, by virtue of being scientists, enjoy "a morally unencumbered freedom from permanent pressure to moral self-reflection" (82). However, role responsibilities of a profession do not generally provide an exemption from general responsibilities. In the years after World War II, Percy Bridgman attempted to provide a more considered argument for such an exemption as controversy swirled over the role of scientists in the construction of the atomic bomb. Bridgman, a Harvard physicist and Nobel laureate, argued, "The challenge to the understanding of nature is a challenge to the utmost capacity in us. In accepting the challenge, man can dare to accept no handicaps. That is the reason that scientific freedom is essential and that the artificial limitations of tools or subject matter are unthinkable" (Bridgman 1947, 153).

The knowledge that scientists produce is so valuable to society, Bridgman suggests, that we must relinquish other claims of social or moral responsibility on scientists so that they can produce this valued end. Scientists, under this view, not only have a need for autonomy (that is, the ability to be the primary decisionmakers in determining the direction of their work), but also have a need to be free from considering the potential consequences of their work beyond the realm of science. The ideas latent in Bridgman's arguments can be cast in two ways, a stronger and a weaker version. The stronger version is that scientific knowledge is valuable beyond price, and thus any sacrifice is worth its achievement. The weaker version is that the price of accepting the burden of moral reflection is too high compared to the value of science.

Is the knowledge produced by scientists so valuable that it is worth the price of scientists' moral exemption from the basic responsibilities articulated above? One way to fashion this claim is to argue that epistemic concerns trump all other values, that is, that the search for truth (or knowledge) is held in such high esteem that all other values are irrelevant before it. If we thought the search for truth (however defined, and even if never attained) was a value in a class by itself, worth all sacrifices, then epistemic concerns alone would be sufficient for considering the consequences of research. The search for truth would overshadow other values, and there would be no need to weigh epistemic concerns against other values. However, there is substantial evidence that we do not accord epistemic concerns such a high status. We place limits on the use of human (and now animal) subjects in research, which indicates we are not willing to sacrifice all for the search for truth.

Such considerations came strongly to the fore after Bridgman wrote his 1947 essay, with the Nuremburg trials and the subsequent concerns over the use and protection of human subjects in scientific research. In addition, our society has struggled to define an appropriate budget for federally funded research, and some high-profile projects (such as the Mohole project in the 1960s[6] and the superconducting supercollider project in the 1990s) have been cut altogether. This suggests that in fact we do weigh epistemic goals against other considerations. That knowledge is important to our society is laudable, but so too is the fact that it is not held transcendently important when compared to social or ethical values. Thus, the knowledge produced by scientists should not be and is not considered priceless. The stronger version of the argument for the moral exemption of scientists fails.

The weaker version of Bridgman's argument is that requiring scientists to consider the consequences of their work (including the consequences of error) is a burden on science and would thus unduly hamper science. In other words, the price of fully morally responsible science is too high. Unfortunately for this line of argument, what that price is has not been articulated or developed. Yet the opposite *can* be clearly argued. One can think of cases where a failure of scientists to consider the unintended consequences, or the implications of error, would be catastrophic. Consider two examples where scientists happily did not view their work as subject to a moral exemption, one from the development of atomic weapons and another from cellular biology. In July 1945, the first atomic bomb was tested in New Mexico. Whatever one may think of the morality of building such weapons, this test of the first plutonium bomb, the Trinity test, was not just a test of a new technology. It was also a decisive test of some of the physical principles that went into the development of the bomb, from the fissibility of plutonium to the calculations behind the implosion device developed by George Kistiakowsky. It was also an experiment about what happens when you produce an explosive chain reaction in the atmosphere. No one had done this before, and there were some worries. One worry that was considered well before the test, and worked on by Hans Bethe, was that the energy in the explosion might produce an explosive chain reaction in the constituents of the earth's atmosphere itself, thus obliterating human life on earth. Happily, the scientists not only thought of this potential outcome, but Bethe pursued the possibility and determined it was scientifically impossible (Rhodes 1986, 419). Only the scientists immersed in the project could have foreseen this and determined the risk was extremely negligible. For a second example, consider the concern scientists raised over recombinant DNA techniques and the resulting Asilomar conference (Culliton 1979, 150–51; Krimsky 1982). Scientists in the midst of exciting research realized that there were risks associated with a new line of research, that serious consequences for public health could occur. They moved to mitigate those risks, even accepting a moratorium on that line of research, thus reducing the likelihood of a damaging unintended result.

In both these cases, scientists, while doing science, reflected on the potential unintended consequences and found the risks unacceptable. Before proceeding with the development of science, they paused and either made sure that the harmful consequences were nearly impossible or figured out ways to make them so. When looking back on these cases, we should be relieved that scientists did not view themselves as morally exempt from considering the risks and consequences of error. The price of morally exempting scientists from the general responsibility to consider the consequences of error looks much higher than the price of having scientists shoulder this burden.

While both of these examples involve scientists considering which testing procedures to follow, or more generally, which methodologies, they do exemplify the difficulties of producing a blanket moral exemption for scientists. The methodological difficulties that might be encountered at the Trinity test were certainly not foreseen at the start of the Manhattan Project. Instead, the concerns with the atmosphere came up in the middle of the project. Given this example, it is clear that we want scientists to think about the potential consequences of error (that is, the potential harms that occur when things do not go as planned) throughout a particular project. Similarly, the recombinant DNA concerns grew out of a set of research projects. Scientists thinking about possible consequences of error in the midst of doing science seems a clearly desirable rather than undesirable aspect of scientific work.

Perhaps those who wish to define a moral exemption for scientists might say that scientists should not consider the consequences of error when they are making general empirical claims, but only when they are considering certain actions they might take that may harm others. However, the distinction between making a general empirical claim and making a judgment about the safety of a scientific process seems dubious at best, showing how difficult it is to keep practical and theoretical reason distinct in this border area. No discussion of the recombinant DNA controversy or the risks of the Trinity test could go forth without making claims about potential risks and the benefits of proceeding. One would sincerely hope that in a case like the Trinity test, scientists would demand a greater burden of proof that an atmospheric chain reaction would not occur than if they were considering the risks of accepting some more mundane claim. Less uncertainty is tolerable when the consequences of error are so high, and in making the claim that the risk of catastrophe is low, we would want a high degree of surety. This *is* the weighing of uncertainty around an empirical claim against the ethical consequences of error. In making crucial empirical claims about the likelihood of error, scientists should consider the consequences of error, for such consideration is what requires different burdens of proof in different contexts. No tenable grounds for a general moral exemption for scientists may be found in a distinction between a judgment regarding the safety of an action and a judgment about an empirical claim.

Why go through these rather desperate attempts to articulate some kind of moral exemption for scientists from a general responsibility we all share? The fear of many proponents of such a moral exemption for scientists seems complex, embedded in a host of considerations. Some are concerned that science will lose its general public authority if a role for social or ethical values is admitted. Yet as chapter 1 demonstrates, no particular piece of science is obviously authoritative anymore, as debates over sound science and junk science rage. The value-free ideal has done nothing to help this debate. Some defenders are concerned that if the current value-free ideal fails, we will revert back to the horrors of science history such as Lysenkoism and Galileo's trial. As I will show in the next chapter, this leap is not warranted. We can reject the current value-free ideal while still holding a constrained role for values in science. Some are concerned that there will be no objectivity left in science if the value-free ideal is gone. Chapter 6 shows that this is not the case, that science can be objective even as it is value-laden. Some are concerned that scientists will be usurping undue authority in our society if the value-free ideal is relinquished. In chapters 7 and 8, I will argue that this need not be the case, particularly as scientists are encouraged to make their judgments explicit. Indeed, the possibility for genuine public input into science increases when the value-free ideal is relinquished.

In sum, there are two important arguments against a moral exemption for scientists. First, there are many cases where such a moral exemption would be very harmful, and no clear boundary can be located that would require moral reflection in some cases and not others. Thus, no blanket exemption is tenable. Second, no convincing argument has been articulated to give scientists even an occasional moral exemption from the consideration of the consequences of their work. With no clear argument for the exemption and a host of considerations against the exemption, I will reject the idea of a general exemption for scientists for the remainder of this work. What are the implications of this rejection for scientists? In the remainder of this chapter, I will discuss the implications for scientists in the advising process. I will also articulate the limits placed on the burden of this responsibility and how the standards of reasonable foresight might work for scientists in practice.

Moral Responsibility and Science Advising

I have argued that scientists can and should bear the burden of the general responsibility to consider the consequences of error. What does the bearing of this burden mean for the role of science in policymaking? As noted in chapter 2, the importance of scientific advice over the past century has been increasing. This advice is used to guide decisions of major social importance. We certainly need scientists in this advisory role, helping to shape and guide decisions with significant technical aspects. We need scientific advice on such issues as which health problems may be posed by air and water pollutants, what options we have for nuclear waste disposal and what risks are associated with them, which drugs should be released on the market and with what labeling, and so forth. What are the moral responsibilities of scientists in this crucial role?

Certainly we should expect honesty and forthrightness from our scientists. To deliberately deceive decisionmakers or the public in an attempt to steer decisions in a particular direction for self-interested reasons is not morally acceptable. Not only would such a course violate the ideals of honesty central to basic science, but it would violate the trust placed in scientists to provide advice, and it would violate the basic ideal of democracy, that an elite few should not subvert the will of the many for their own gain.

But whether scientists should be honest is not the dispute at the heart of the value-free ideal for science. The issue is whether scientists should consider the consequences of error in their advising, including errors that may lead to harm. Should scientists worry about being reckless or negligent when giving advice; that is, should they consider the consequences of error when deciding how much evidence is enough to support making an empirical claim? As noted in chapter 3, Rudner developed this line of argument in the 1950s, suggesting that scientists should consider not just the extent of the uncertainty inherent in any scientific statement, but should also weigh the importance of the uncertainty by considering the consequences of error. Such a weighing would require the use of social and ethical values in scientific judgments. So, the issue is whether or not scientists, when placed in official advisory positions or simply providing the general public with authoritative advice, should consider the consequences of possible errors in their advice.

If my arguments in the preceding section are convincing, we certainly would want scientists to consider the potential consequences of error when giving advice. Scientists have the same obligations as the rest of us not to be reckless or negligent, and this obligation also holds when a scientist is making an empirical claim (the basic component of advising). This means that when a scientist makes an empirical claim in the process of advising, they should consider the potential consequences if that claim is incorrect. In the advising context, this includes possible social and ethical consequences of policymakers acting on the basis of the empirical claim. The scientist acting as an advisor should consider the extent of uncertainties around the claim and the possible consequences of incorrectly accepting or rejecting the claim, and they should weigh the importance of the uncertainties accordingly. Thus, science advising should not be value free.

Note, however, that social and ethical values can legitimately enter into the advice only through the weighing of uncertainty. The scientist should not think about the potential consequences of making an accurate empirical claim and slant their advice accordingly. Only in the weighing of uncertainty do social and ethical values have a legitimate role to play when deciding, based on the available evidence, which empirical claims to make.

An example will help to clarify this ideal for fully responsible science advice. Suppose a scientist is examining epidemiological records in conjunction with air quality standards and the scientist notices that a particular pollutant is always conjoined with a spike in respiratory deaths. Suppose that this pollutant is cheap to control or eliminate (a new and simple technology has just been developed). Should the scientist make the empirical claim (or, if on a science advisory panel reviewing this evidence, support the claim) that this pollutant is a public health threat? Certainly, there is uncertainty in the empirical evidence here. Epidemiological records are always fraught with problems of reliability, and indeed, we have only a correlation between the pollutant and the health effect. The scientist, in being honest, should undoubtedly acknowledge these uncertainties. To pretend to certainty on such evidence would be dishonest and deceptive. But the scientist can also choose whether or not to emphasize the importance of the uncertainties. And this is where the weighing of the consequences of error comes in. If the scientist accepts that claim as sufficiently reliable (not certain) and is wrong, little expense will have been accrued as policymakers act on the scientist's advice. If the scientist rejects the claim as insufficiently reliable or well supported and is wrong, public health will be damaged substantially. In this admittedly easy case, the fully responsible advice would be to note the uncertainties but, on the basis of the consequences of error, suggest that the evidence available sufficiently supports the claim that the pollutant contributes to respiratory failure. Such advice would be fully morally responsible, and not value free.

One might argue that considering the consequences of error when giving advice on an advisory panel or some other formal advising mechanism is acceptable, and indeed a responsibility of the participants in such a mechanism. However, one may still want to claim that scientists in general should eschew the considerations of such consequences in other forums when making empirical claims. In other words, one may want to hold to a distinction between the scientist qua scientist and the scientist qua advisor, accepting the need for values in the latter role while rejecting them in the former.

The attempt to draw a distinction in the practices of scientists between the two roles, scientist and advisor, is dubious at best. The scientist is called on to be an advisor because she is a scientist, and the advice is to be based on her expertise as such. But the problem runs even deeper. Scientists hold a broadly authoritative position in our society, regardless of whether they are functioning in a formal advising role or not. Thus, when scientists make empirical claims, whether in scientific conferences, in science journals, or on an advisory panel, those empirical claims carry with them a prima facie authority. This is why science journalists are interested in science conferences, why scientific journals are covered in the general press, and why scientists are asked for their views in areas related to their expertise. This basic authority is what turns a general responsibility to consider the consequences of error into a particularly important responsibility for scientists. Because their empirical claims carry with them this prima facie authority, the potential consequences of error can be more far-reaching than for a nonscientist. If I make a claim that a plant is a dangerous invasive species that can be eradicated, no one is likely to listen or act on the basis of my claim. There are few consequences of error for me to consider. But an ecologist making such a claim is far more likely to be believed, and to have their claim acted upon. The authority of science in society makes a distinction between scientist qua scientist and scientist qua advisor untenable. It also can place a heavy burden on scientists to meet their moral responsibilities fully. It is time to reflect upon the limits of this burden, not just the full extent of its reach.

Limits to the Moral Responsibilities of Scientists

The positions of Bridgman and Lübbe are untenable. Simply because scientists provide us with important knowledge cannot and does not exempt them from basic moral responsibilities, which include reflecting on the implications of their work and the possible consequences of error. Yet scientists cannot be responsible for every use or misuse of their work. Fortuitously, the basic understanding of moral responsibility articulated in the beginning of this chapter provides guidelines for when and how much moral responsibility scientists must bear for their work and their advice.

As noted above, when considering the consequences of error, a person is not responsible for every consequence that follows from their action. Rather, a person is held responsible only for those consequences that are reasonably foreseeable. Thus, in our earlier example of setting a field ablaze to clear brush, it is easily foreseeable that on a hot, windy day the fire could get out of control and burn a neighbor's property. To proceed with a burn under such conditions is reckless (if one sees the risk but does not care) or negligent (if one fails to foresee the risk). We can demand such foresight in these cases because any reasonable person would be able to foresee the risks from such an action.

Scientists should be held to a similar standard of foresight, but indexed to the scientific community rather than the general public. Because scientists work in such communities, in near constant communication and competition with other scientists, what is foreseeable and what is not can be readily determined. As with other ideas in science, potential consequences of error spread quickly, and scientists discuss pitfalls, dangers, and uncertainties readily. Another example from nuclear physics shows the ready benchmark of foreseeability that can exist in science. Throughout the 1930s, after the discovery of the neutron by James Chadwick, nuclear physics blossomed. The neutron provided a ready tool for the probing of the nucleus, and many discoveries followed. However, none of these discoveries seemed to have any important implications outside of nuclear physics. The idea that usable energy may be derived from nuclear processes was thought to be unfounded speculation, or "moonshine," as Ernest Rutherford put it. All this changed in December 1938 with the discovery of fission. No one had thought that an atom might actually split into two large chunks before Lise Meitner's insight into Otto Hahn and Fritz Strassmann's experiments. There was no foreseeability of the usefulness of nuclear processes for a bomb or useful energy production until this point.[7] As word of fission crossed the Atlantic in January 1939, it was clear to all what fission meant: the possibility for useful nuclear energy, either as a power source or a bomb.[8] In the political climate of the time, this worried many, and debate soon followed on how to proceed. But the foreseeability of nuclear weapons can be pinpointed to this moment in time. It would be absurd to have expected Chadwick to have foreseen this when he discovered the neutron in 1932. By the spring of 1939, few nuclear physicists had not foreseen the disturbing potential of fission.[9] The nuclear physics community provides a ready benchmark for what a reasonable scientist could foresee in this case.

The example of fission and the sudden usefulness of nuclear physics is an example of the foreseeability of the direct consequences of a line of research. Much of this chapter is about the foreseeability of error and its consequences. Here, too, scientific communities provide ready benchmarks. The concern of scientists working with recombinant DNA over the potential consequences of laboratory error, particularly the accidental generation and release of new biohazards, led directly to the moratorium on such research by 1974. Scientists in the field could readily foresee the dangers and worked together to forestall them. And the consequences of error in making empirical claims were also generally agreed upon and foreseeable. Today, it is precisely the concern over these consequences that drives so many of our more contentious technical debates. For example, if a chemical is known to cause cancer in humans, a regulatory response can be predicted. So making the empirical claim that a chemical causes cancer brings with it clear consequences of error, namely unnecessary regulation (unnecessary if the claim is erroneous). On the other hand, not making the empirical claim (perhaps because one suspects the chemical but does not think the evidence sufficient) also carries with it clear consequences of error, namely the cancer will continue to be caused. The consequences of error are readily foreseeable by all, and are often a central engine of the debate.

Requiring that scientists consider the consequences of their work does not mean requiring that they have perfect foresight. The unexpected and unforeseen can and does happen. Holding scientists responsible for unforeseen consequences is unreasonable. What is reasonable is to expect scientists to meet basic standards of consideration and foresight that any person would share, with the reasonable expectations of foresight judged against the scientist's peers in the scientific community. Thus, the moral burdens on scientists are not unlimited. They are held to only what can be foreseen, and thus discussed and considered.

It must also be noted here that scientists need not carry this burden alone. As mentioned above, scientists already use the assistance of internal review boards to help them meet their responsibilities with respect to methodologies involving human and (some) animal subjects. However, if scientists found it useful, they might consider convening similar kinds of bodies, either permanent or temporary, that could help them make difficult choices when they arise. I will discuss some of these possible mechanisms in the final chapter. There are many interesting ways in which scientists can shift the burden of reflecting on these issues to others. However, the scientist can never abdicate the responsibility completely. Often only the scientists on the cutting edge will fully understand the implications and the risks of their work. We will always need them to reflect on those as they proceed into the unknown.

Conclusion

We all share a general responsibility to consider the consequences of our choices, including the consequences of error. This responsibility extends to the making of empirical claims, an activity central to any examination of science in public policy. Because scientists have no good reason to be exempt from this general responsibility, and indeed we have good reason to want them to shoulder (at least in part) the responsibility, scientists must weigh the consequences of error in their work. The values needed to weigh those consequences, and thus determine the importance of uncertainties in science, become a required part of scientific reasoning. The value-free ideal for science can no longer be held up as an ideal for scientists.

Some might argue at this point that scientists should just be clear about uncertainties and all this need for moral judgment will go away, thus preserving the value-free ideal. It is worth recalling Rudner's response to a similar argument from Jeffrey discussed in the previous chapter. Even a statement of uncertainty surrounding an empirical claim contains a weighing of second-order uncertainty, that is, whether the assessment of uncertainty is sufficiently accurate. It might seem that the uncertainty about the uncertainty estimate is not important. But we must keep in mind that the judgment that some uncertainty is not important is always a moral judgment. It is a judgment that there are no important consequences of error, or that the uncertainty is so small that even important consequences of error are not worth worrying about.[10] Having clear assessments of uncertainty is always helpful, but the scientist must still decide that the assessment is sufficiently accurate, and thus the need for values is not eliminable.

The demise of the value-free ideal may be disturbing to some. What is the proper role for values in science, if science is not value free? Can values, any values, play any role whatsoever in science? Can they dictate a scientific result? Is the objectivity of science, the source of its authority, doomed if the value-free ideal is rejected? I will take up these difficult questions in the next two chapters and argue that values should play only a constrained role in scientific reasoning. Thus, the demise of the value-free ideal does not mean values can run roughshod over evidence and reasoning. Science can be objective while remaining value saturated.

CHAPTER 5

THE STRUCTURE OF VALUES IN SCIENCE

EVEN WHEN MAKING EMPIRICAL CLAIMS, scientists have the same moral responsibilities as the general population to consider the consequences of error. This apparently unremarkable statement has some remarkable implications. It means that scientists should consider the potential social and ethical consequences of error in their work, that they should weigh the importance of those consequences, and that they should set burdens of proof accordingly. Social and ethical values are needed to make these judgments, not just as a matter of an accurate description of scientific practice, but as part of an ideal for scientific reasoning. Thus, the value-free ideal for science is a bad ideal. However, simply discarding the ideal is insufficient. Although scientists need to consider values when doing science, there must be constraints on how values are considered, on what role they play in the reasoning process. For example, simply because a scientist values (or would prefer) a particular outcome of a study does not mean the scientist's preference should be taken as a reason in itself to accept the outcome. Values are not evidence; wishing does not make it so. There must be some important limits to the roles values play in science.

To find these limits, it is time to explore and map the territory of values in science. This will allow me to articulate a new ideal for values in science, a revised understanding of how values *should* play a role in science and of what the structure of values in science *should* be. I will argue that in general there are two roles for values in scientific reasoning: a direct role and an indirect role. The distinction between these two roles is crucial. While values can play an indirect role throughout the scientific process, values should play a direct role only for certain kinds of decisions in science. This distinction between direct and indirect roles allows for a better understanding of the place of values in science—values of any kind, whether cognitive, ethical, or social. The crucial normative boundary is to be found not among the kinds of values scientists should or should not consider (as the traditional value-free ideal holds), but among the particular roles for values in the reasoning process. The new ideal that rests on this distinction in roles holds for all kinds of scientific reasoning, not just science in the policy process, although the practical import of the ideal may be most pronounced for policy-relevant science.

In order to map the values in science terrain, we need to consider the function of values throughout the scientific research process. The schema of the research process I use in this chapter is admittedly idealized, but it should be both familiar and a sufficient approximation. The first step in any research endeavor is deciding which questions to pursue, which research problems to tackle. This decision ranges from the rather vague ("I think I'll look here") to the very precise (a particular approach to a very well-defined problem). Regardless, a decision, a judgment, must be made to get the process started. Then the researcher must select a particular methodology in order to tackle the problem. This decision is often closely tied to the choice to pursue a particular problem, but in many cases it is not. These decisions are often under constraints of ethical acceptability, resource limitations, and skill sets. And if institutional review boards are overseeing methodological choices, these decisions can take place over a protracted period of time. Methodological choices also profoundly shape where one looks for evidence.

Once the researcher embarks upon a chosen methodology, he or she must decide how to interpret events in the study in order to record them as data. In many cases, this is a straightforward decision with little need for judgment. However, judgment may be called for on whether a particular event occurred within the framework of the methodological protocol (was the equipment working properly?), or judgment may be needed in how to best characterize an ambiguous event. Once the researcher has collected the data, he or she must interpret it and draw conclusions. The scientist must ultimately decide whether the data support the hypothesis, and whether to accept or reject a theory based on the evidence. This process is mimicked in the papers published by scientists, and thus provides a useful heuristic for understanding the scientific process. These decision points may not be so orderly or neat in actual practice, but we can generally recognize where in actual practice a decision made by a scientist would be situated in this idealized process. I will use this schema of the research process to explicate where values should play which kind of role.

Before delving further into the *roles* values play in science, we should first examine whether clear distinctions can be made among the *kinds* of values potentially relevant to science. The value-free ideal rests on at least one distinction for values in science, between acceptable and unacceptable values. Acceptable values became known as "epistemic," meaning related to knowledge, whereas the rest became known as "nonepistemic," a catch-all category that includes social, ethical, and other values—all the "forbidden" values. If this distinction fails, not only is there another reason to reject the value-free ideal (for one of its foundational distinctions is faulty), but also one should not rely upon that distinction in forging a new ideal.

In recent examinations of values in science, the divisions have become more complex, and, upon closer examination, simple distinctions are less tenable. If we cannot rely upon definitively demarcated categories of values, it becomes all the more important that we keep clear the *role* of the values in our empirical reasoning, making sure that they are constrained to legitimate roles. In addition, without the dichotomy between epistemic and nonepistemic values, we can better understand the tensions involved in weighing various kinds of values in any given scientific judgment, for example, when cognitive and ethical values conflict. A topography of values, with continuities among categories rather than cleanly demarcated categories, can help to organize the kinds of values involved.

The Topology of Values in Science

Central to the current value-free ideal, the 1950s debate about values in science (discussed in chapter 3) introduced a demarcation in the types of values that could influence science. "Epistemic values" were thought to be those that related directly to knowledge and could be differentiated from nonepistemic values such as social and ethical values. Traditional examples of epistemic values included predictive accuracy, explanatory power, scope, simplicity (or "economy"), and so forth.[1] Under the value-free ideal, these values were sharply demarcated from social and ethical values, such as concern for human life, reduction of suffering, political freedoms, and social mores.

The clear demarcation between epistemic (acceptable) and nonepistemic (unacceptable) values is crucial for the value-free ideal. Philosophers of science, even while presuming the isolation of science from society, have understood the endemic need for judgment in science. Scientists often disagree over which theory is preferable in accounting for the same available evidence. So even with an isolated science, there is a need for some value judgments in science. Philosophers have thus attempted to make a distinction between the kinds of values proper to scientific judgment and the kinds of values thought to threaten the value-free ideal. As Levi (1960, 1962) argued, and most philosophers came to agree, only values that were part of the "canon of scientific reasoning" should be used, that is, only epistemic values were legitimate when assessing the strength of evidence in relation to theory. However, the reasons for such an isolated science were neither fully articulated nor well supported. Indeed, as I argue in chapter 4, such scientific isolation from moral responsibility was unwarranted and undesirable. Even so, it is not clear that the initial distinction between epistemic and nonepistemic values was a viable one.[2] If this distinction fails, the value-free ideal is untenable, the arguments of the previous chapter aside.

The main argument for the porousness (and thus failure) of the epistemic/nonepistemic distinction is that epistemic values end up reflecting the nonepistemic values of the day. For example, Phyllis Rooney (1992, 16) observes that Ernan McMullin, a staunch defender of the distinction, allowed an acceptable role for "nonstandard epistemic values," such as metaphysical presuppositions or theological beliefs. An example of this, Rooney notes, would be the theological views on the role of randomness in the universe that underlay the Bohr-Einstein debate. The social or cultural values that shaped Bohr's or Einstein's theological views then acted as guides for epistemic choice, thus operating as epistemic values. Such nonstandard epistemic values have deeply influenced the direction of scientific thought, but they also often reflected social values, Rooney suggests, thus smuggling the nonepistemic values through the boundary. Similar issues arise in scientific studies of gender, where socially desired biological determinism appears within scientific research in the form of simplicity, a supposedly acceptable and pure epistemic value (Rooney 1992, 18). Thus, the "'non-epistemic' [becomes] encoded into the constitutive [that is, epistemic] features of specific theories" (ibid., 20). The boundary looks quite permeable in this light, unable to bear the weight of keeping the undesirable values out of science.

Helen Longino (1996) makes a similar point. She first provides examinations of the standard epistemic values, such as accuracy, consistency, simplicity, scope, and fruitfulness (41–44). Then in contrast, she suggests some alternative epistemic virtues arising from feminist critiques of science, such as novelty, applicability, and ontological heterogeneity (45–50). With two competing sets of epistemic values in hand, Longino both undermines the apparent political neutrality of the standard set of epistemic values and raises the question of whether a clear dichotomy can be made between internal acceptable and external unacceptable values.

Even Hugh Lacey, another defender of the acceptable versus unacceptable values in science distinction, has difficulty maintaining a firm boundary. In his discussion of Longino's work, he writes,

> Longino maintains that a cognitive [that is, epistemic] value such as empirical adequacy, does not have "a solely epistemic or cognitive basis." I concur: since adopting a strategy is partly rationalized in view of its mutually reinforcing interactions with certain social values, values contribute to some extent to the interpretation of empirical adequacy that one brings to bear on one's hypotheses. (Lacey 1999, 221)

How one can maintain the dichotomy of values needed to support the traditional value-free theses is unclear in the face of these concerns. If social (nonepistemic) values shape the instantiation of epistemic values, then social values are influencing science through epistemic values.[3] Under the burden of these arguments, the strict demarcation between epistemic and nonepistemic values no longer seems viable. Without such a demarcation, the dichotomy between acceptable and unacceptable values in science fails. In the absence of such a dichotomy, a more nuanced topography of values in science is both possible and desirable.

In developing such a topography, we should begin by examining the goals behind the stated values. With a clearer understanding of goals, we can have a clearer understanding of the types of values relevant to science. Having types of values does not commit one to strict demarcations, however. It is more useful to think about these values as spread out in a landscape, with different parts of the landscape focusing on different goals, rather than as segregated into camps. Areas of overlap and interaction are common. For example, how one interprets a cognitive value such as simplicity may well be influenced by socially structured aesthetic values, values that help one interpret and recognize simplicity. Or which theories seem more fruitful may have much to do with which lines of investigation are likely to receive the needed social resources. When one rejects the idea that there are kinds of values acceptable and unacceptable in science, the need for clearly differentiated categories dissolves, and interactions or tensions can be more readily noticed.

With such a fluid understanding of the categories, how should we understand the topography of values in science? What are the various goals relevant to science? For the remainder of this work, three categories of values will reflect the predominant concerns: ethical, social, and cognitive. (Aesthetic values are often also important in some areas of science, but tend to be less so for policy-relevant science.) I will argue that epistemic values, *distinguished from cognitive values*, should not be thought of as values at all, but rather as basic criteria that any scientific work must meet. The epistemic criteria are set by the very valuing of science itself, and thus establish the scope of science, rather than acting as values within science.

The first kind of value, ethical value, focuses on the good or the right. These values are particularly important when examining the consequences of error for the general public. Ethical values help us weigh whether potential benefits are worth potential harms, whether some harms are worth no price, and whether some harms are more egregious than others.[4] Examples of ethical values relevant to scientific research include the rights of human beings not to be used for experimentation without fully informed consent, the consideration of sentient beings for their pain, concern for the death and suffering of others, whether it is right to pursue research for new weapons of mass destruction, and whether an imposition of risk is ethically acceptable.

Closely related (but not identical) to ethical values are social values. Social values arise from what a particular society values, such as justice, privacy, freedom, social stability, or innovation. Often social values will overlap with ethical values. For example, the social concern one might express over poverty can be tied to issues of justice or to concern over the increased health risks borne by impoverished individuals. A scientist may research a particular area because of both an ethical concern and a social value that reinforces that concern, such as pursuing research on malaria because it is a dreadful disease affecting millions and the scientist wishes to reduce the suffering caused by it, *and* because the scientist shares the social concern over the justice of working on diseases that afflict an affluent few rather than diseases that cause excessive suffering among the impoverished many (Flory and Kitcher 2004). Nevertheless, some social values can be opposed to ethical values. For example, the social value of stability was antithetical to the ethical values underlying the push for desegregation and the civil rights movement. More recently, the social value of stability and the reinforcement of stereotypes undergirding that stability, are reflected in some scientific studies of race and IQ that seek to show that the current social order is naturally ordained in some way (Fancher 1985; Gould 1981). Such values run directly counter to ethical values focused on the rights and qualities of individuals as such.

Cognitive values make up another major grouping. By "cognitive values" I mean something more precise than the vague clumping of acceptable values in science central to the value-free ideal, often equated with epistemic values. Rather, I mean those aspects of scientific work that help one think through the evidential and inferential aspects of one's theories and data. Taking the label "cognitive" seriously, cognitive values embody the goal of assisting scientists with their cognition in science.

For example, simplicity is a cognitive value because complex theories are more difficult to work with, and the full implications of complex theories are harder to unpack. Explanatory power is a cognitive value because theories with more explanatory power have more implications than ones that do not, and thus lead to more avenues for further testing and exploration. (Explanatory theories structure our thinking in particular but clearly articulable ways, and this allows one to draw additional implications more readily.) Scope is a cognitive value because theories with broad scope apply to more empirical areas, thus helping scientists develop more avenues for testing the theories. The consistency of a theory with other areas of science is a cognitive value because theories consistent with other scientific work are also easier to use, allowing for applications or extensions of both new and old theories, thus again furthering new research. Predictive precision is a cognitive value because making predictions with precision and testing to see if they are accurate helps scientists hone and refine theories more readily.[5] And fruitfulness is an obvious cognitive value because a productive theory provides scientists with many avenues for future investigation. Fruitfulness broadly construed may be considered the embodiment of cognitive values—the presence of any cognitive value should improve the productivity of an area of science. It should allow for more predictions, new avenues of testing, expansion of theoretical implications, and new lines of research. In sum, cognitive values are concerned with the possibilities of scientific work in the immediate future.

If cognitive values are about the fruitfulness of research, epistemic values are about the ultimate goal of research, which is true (or at least reliable) knowledge. As Laudan (2004) points out, many cognitive values have little to do with truth or truth preservation. Laudan argues that virtues such as scope, generality, and explanatory power "are not epistemic virtues," as they have no necessary connection to whether a statement is true (18). Simply because one statement explains more than another does not mean the latter statement is false. It might, however, be less useful to scientists, lacking sufficient cognitive value. In this light, values remaining in the epistemic category are quite limited. Internal consistency should be considered an epistemic value, in that an internally inconsistent theory must have something wrong within it. Because internal inconsistency implies a fundamental contradiction within a theory, and from a clear contradiction any random conclusions (or predictions) can be drawn, lacking internal consistency is a serious epistemic failing. In addition to internal consistency, predictive competence should be considered an epistemic value. If a theory makes predictions that obviously fail to come to pass, we have serious reason to doubt the theory. Predictive competency should be thought of in a minimal way, similar to empirical adequacy[6] or conforming to the world.[7] Predictive competency is thus not the same as predictive precision. A theory can have predictive competency and still not be terribly precise in its predictions, nor terrifically accurate in its success. Competency just implies "close enough" to remain plausible.

These epistemic virtues operate in a negative way, excluding claims or theories that do not embody them, rather than as values, which are aspects of science for which to strive, but which need not be fully present in all cases. For this reason, so-called "epistemic values" are less like values and more like criteria that all theories must succeed in meeting. One must have internally consistent theories; one must have empirically adequate/conforming/predictively competent theories. Without meeting these criteria, one does not have acceptable science.[8] This claim follows from the reason we value science in the first place, that science is an enterprise that produces reliable knowledge. This reliable knowledge is valued for the guidance it can provide for our understanding and our decisions. A predictively incompetent theory is clearly not reliable; an internally inconsistent theory can give no guidance about the world. The goal of science—reliable knowledge about the world—cannot be achieved without meeting these criteria. The other values discussed above, the ethical, the social, and the cognitive, serve different goals and thus perform a different function in science, providing guidance at points of judgment when *doing* science, helping one weigh options. Epistemic criteria determine what the viable options are among scientific theories, acting as baseline requirements that cast serious doubt on the acceptability of a scientific theory when they are not met.

Once we exclude epistemic criteria from our consideration of values in science, it is easier to see how values should play a role in scientific reasoning. Values can be used to make judgments at several places in the scientific process. There are different kinds of choices to be made as the scientist moves through the process, and at different points, different roles for values are normatively acceptable. All the kinds of values described here can have a relevant and legitimate role throughout the process, but it is the *role* of the values in particular decisions that is crucial.

The Structure of Values in Science: Direct and Indirect Roles

Multiple kinds of values are needed throughout science, contrary to the value-free ideal. Depending on the nature of the judgment required, there are acceptable and unacceptable roles for values in science. Preventing the unacceptable roles preserves the integrity of science (regardless of whether the science is policy relevant or not) while allowing science to take its full measure of responsibility for its prominent place in public decisionmaking (in the cases where the science is policy relevant).

To set the scope of our discussion, let us first presume that we value science. The very decision to pursue scientific inquiry involves a value-laden judgment, namely that such an inquiry is a worthwhile pursuit. The persons deciding to invest their time and resources into science must believe that doing science is a valuable enterprise. What kind of value underlies this decision? It is neither a truth-preserving epistemic criteria nor a cognitive value assisting specific choices. Rather, it is a social value, for it reflects the decision of the society as a whole that science is a valuable and desirable pursuit, that we want to have a more accurate and complete understanding of the way the world is, and that science is a good way to get that understanding. It need not be this way. We could have a society that values stability above all else and cares little for this kind of understanding, with its constant revisions and changes. We could have a society that values stasis above truth and thus eschews the pursuit of changeable knowledge. Happily, from my perspective, we do not. The very decision to pursue science thus depends upon a social value.

Because of the value we place on science, because we care about having a reliable understanding of the world, science must meet certain standards to be acceptable. This sets up the epistemic criteria discussed above. It is because we care about having reliable empirical knowledge that scientific theories must be internally consistent and predictively competent. Our discussion of values in science must take place against the backdrop of this value judgment—the valuing of science itself as the source for reliable empirical knowledge.

There are two kinds of roles for values *in* science—direct and indirect. To see the difference between the two roles, let us briefly consider the decision context most central to scientists and philosophers, the decision of whether to accept or reject a theory based on the available evidence. Two clear roles for values in reasoning appear here, one legitimate and one not. The values can act as reasons in themselves to accept a claim, providing direct motivation for the adoption of a theory. Or, the values can act to weigh the importance of uncertainty about the claim, helping to decide what should count as *sufficient* evidence for the claim. In the first direct role, the values act much the same way as evidence normally does, providing warrant or reasons to accept a claim. In the second, indirect role, the values do not compete with or supplant evidence, but rather determine the importance of the inductive gaps left by the evidence. More evidence usually makes the values less important in this indirect role, as uncertainty reduces. Where uncertainty remains, the values help the scientist decide whether the uncertainty is acceptable, either by weighing the consequences of an erroneous choice or by estimating the likelihood that an erroneous choice would linger undetected. As I will argue below, a direct role for values at this point in the scientific process is unacceptable, but an indirect role is legitimate.

The difference between a direct role for values in science and an indirect role is central to our understanding of values in reasoning throughout the scientific process. In the *direct* role, values determine our decisions in and of themselves, acting as stand-alone reasons to motivate our choices. They do this by placing value on some intended option or outcome, whether it is to valorize the choice or condemn it. The value provides warrant or reason, in itself, to either accept or reject the option. In this direct role, uncertainty is irrelevant to the importance of the value in the judgment. The issue is not whether the choice will somehow come out wrong in the end, but whether the choice, if it comes out as expected, is what we want. This role for values in science is crucial for some decisions, but we will see that it must be restricted to certain decisions made in science and excluded from others. The integrity of the scientific process cannot tolerate a direct role for values throughout that process.[9]

The *indirect* role, in contrast, can completely saturate science, without threat to the integrity of science. This role arises when there are decisions to be made but the evidence or reasons on which to make the decision are incomplete, as they so often are, and thus there is uncertainty regarding the decision. Then values serve a crucial role of helping us determine whether the available evidence is sufficient for the choice and what the importance of the uncertainty is, weighing the potential consequences of a wrong choice and helping to mitigate against this possibility by requiring more evidence when such consequences are dire. But the values in this role do not determine the choice on their own. If we find new evidence, which reduces the uncertainties, the importance of the relevant value(s) diminishes. In this indirect role, more evidential reasons in support of a choice undercut the potency of the value consideration, as uncertainty is reduced. The value only serves as a reason to accept or reject the current level of uncertainty, or to make the judgment that the evidence is sufficient in support of a choice, not as a reason to accept or reject the options per se.

This distinction between the direct and indirect role for values can be seen in Heil 1983. Heil points out that motives or incentives for accepting empirical claims should not be conflated with reasons that provide "epistemic *support*" for those claims (755). In other words, while values may provide a motivation to believe a claim, values should not be construed as providing epistemic support for a claim. Values are not the same kind of thing as evidence, and thus should not play the role of providing warrant for a claim. Yet we can and do have legitimate motives for shifting the level of what counts as *sufficient* warrant for an empirical claim. Even as Heil warns against the self-deceptive, motive-driven use of values in support of empirical claims, he notes,

> There are many ways in which moral or prudential considerations can play a role in the factual conclusions one is likely to draw or the scientific pronouncements one may feel entitled to issue. We regard it as reasonable to require that a drug possessing an undetermined potential for harm pass rather more stringent tests than, for example, a new variety of hair tonic, before it is unleashed on the public. In such cases, however, we seem not to be tampering with ordinary evidential norms, but simply exercising reasonable caution in issuing judgments of safety. (761)

A footnote in Heil's essay elaborates this point: "One checks the oil and water in one's automobile with special care before setting out on a long journey. This is not because one needs more evidence about one's automobile in such cases than in cases in which one is merely driving across town. It is rather that there is a vast difference in the consequences of one's being wrong" (761n12) The indirect role is on display here, as the consequences of error come to the fore in the indirect role. Thus, the epistemic agent is concerned with whether the evidence is sufficient for supporting a claim that if wrong would prove a deleterious basis for further decisions (like deciding to drive one's car on a long journey or marketing a new drug).

In what particular instances do values legitimately play their distinct roles? First, there are important limits on the direct role for values in science. There are many choices made in science, particularly at the heart of the scientific process, where no direct role for values is acceptable. If we do indeed care about gaining reliable knowledge of the world, it is crucial to keep these limits on values in science. Second, these limits on the role of values in science hold regardless of whether the values we are considering are ethical, social, or cognitive. All of these values can have direct or indirect roles in science, and all must keep to the limits I will describe below. Thus, maintaining a distinction in the *kinds* of values to be used in science is far less important than maintaining a distinction in the *roles* those values play.

Direct Roles for Values in Science

There are several points in the scientific process where a direct role for values is needed. At these points, values can and should direct our choice, telling us which option should be pursued. Values serve an important function in the direct role, particularly for the early stages of a research project, one that should not be overlooked in discussions of the nature of science. In fact, they have been increasingly acknowledged over the past fifty years, as concerns over which projects to pursue and how to ethically utilize human subjects have increased. The proper direct role for values in science has never been a source of worry for philosophers of science, even for those who hold to the value-free ideal.[10]

The first point for a direct role for values is in the decision of which scientific projects to undertake. There are a far larger number of possible scientific studies than can actually be pursued. How we select which ones to pursue is a complex process, a combination of which projects scientists deem feasible and interesting, which projects government and/or private interests think are worth funding, and which projects are ethically acceptable. In these decisions, values play a direct role in shaping scientists' choices, the values serving to weigh the options before the scientist. A scientist may wish to study the migration patterns of whales, for example, because he or she cares about the whales and wants to understand them better in order to improve protections for them. The value (perhaps ethical, perhaps social) the scientist places on the whales is the reason for the scientist's interest, directing the choice. Another scientist may pursue a particular study of a chemical substance because of what that study may reveal about the nature of chemical bonding, a topic pursued purely for the interest in the subject irrespective of any applications. The value scientists place in their intellectual interests is sometimes the primary reason for the choice.

Funding decisions in science also embody the direct role of values. We choose to fund areas of research about which we care to know more. [SOCIAL] The government may fund a project studying the possibility of increased photovoltaic cell efficiency because the grant administrators consider any increase in efficiency to be important to the country's future energy security and economic development. The social value the administrators place on energy security and economic development directs the kinds of projects they fund. Other values more central to cognitive concerns may drive the funding of other studies, to increase the scope of a particular area of study or to test the predictive precision of a new theory. Thus, ethical, social, and cognitive values help scientists decide where to direct their efforts and are reflected in both funding decisions and the scientists' own choices.

Once one has selected a particular area of research and which question to attempt to answer, one must decide which methodology to pursue, and here concerns over the ethical acceptability of a methodology will directly shape scientific decisions. [ETHICAL] If the chosen methodological approach involves flatly unethical actions or treatment of others, the approach should be rejected. The ethical values can legitimately direct the selection of methodologies, particularly when human subjects are involved. For example, a scientist may wish to use human subjects in a physiological study that would expose those subjects to substantial risk of injury. The internal review board for the project may decide that the methodology as laid out by the scientist is unethical, and require the scientist to change the approach to the subjects of the study, or perhaps even to scrap the study altogether. Here an ethical value (a concern over the treatment of human subjects) directs the way in which a study will be conducted, and even if it can be conducted.

There can be conflicts among values for these choices about methodology. Consider this admittedly extreme example. It would be very evidentially useful to test pesticides on humans, particularly a large pool of humans in controlled conditions over a long period of time. However, despite the cognitive advantages of such a testing procedure (including the possibility for precise testing of a theory), there are some serious ethical and social concerns that would recommend against this testing procedure. It is very doubtful one could obtain adequate volunteers, and so coercion of subjects would be useful for the study, but ethically abhorrent. Both social and ethical values stand against such a study, for it is doubtful society would approve of such a testing regimen, and very likely a social outrage against the scientists would be expressed when word of the study got out. (An outcry was heard recently against a much less controlling study involving following the exposures to and health of children in families using pesticides in Florida. Imagine if the experimental subjects were exposed against their will.)[11] However, such a test would provide far greater precision to predictions about the health effects of pesticides. So, despite the cognitive value of such a test, the conflicting ethical and social values would overrule that value.

In these examples, the relevant values (social, ethical, and cognitive) act in the direct role, providing reasons for the choices made. If the methodology proposed is unethical, it cannot be pursued by the scientist. If the project proposed is not desired by funders, it will not receive funding support. If a project is not interesting to scientists, they will not pursue it.[12] Conversely, a more ethical methodology, a socially desired research program, a cognitively interesting project, will be selected because of the value placed on those options. The values are used to weigh the options available, determining whether they are acceptable and which is preferable.

Not all direct roles for values in these early stages of science should be acceptable, however. One cannot use values to direct the selection of a problem and a formulation of a methodology that in combination predetermines (or substantially restricts) the outcome of a study. Such an approach undermines the core value of science—to produce reliable knowledge—which requires the possibility that the evidence produced could come out against one's favored theory. For example, suppose a scientist is studying hormonal influences on behavior in children.[13] It is already known that (a) there are hormonal differences in children, and (b) there are behavioral differences in children. A study that simply measures these two differences to find a correlation would be inadequate for several reasons. First, a mere correlation between behavior and hormones tells us little about causation, as we also know that behavior can change hormone levels. Second, we know there are other important factors in behavior besides hormones, such as social expectations. A study that merely examines this one correlative relationship only tells us something interesting against a backdrop of a presumption that hormones determine behavior. If one assumes this going in, deliberately structures the study so as to exclude examination of other possibilities, and then claims the results show that hormones determine behavior, values have played an improper direct role in the selection of research area and methodology. For by valuing the support of the scientist's presumptions in developing the project, the scientist has distorted the very thing we value in science—that it can revise itself in the face of evidence that goes against dearly held theories. By making choices that preclude this, the scientist has (deliberately or inadvertently) undermined a crucial aspect of the scientific endeavor.

Therefore, a direct role for values in the early stages of science must be handled with care. Scientists must be careful that their methodology is such that it can genuinely address the problem on which they have chosen to focus, and that they have not overly determined the outcome of their research. Research by feminist scholars of science has made clear how difficult this can be for individual scientists—that often the underlying presumptions that structure an approach (and may well predetermine the outcome) are opaque to the individual scientist. It is for this reason that calls for diversity in scientific communities are so important, for having colleagues coming from diverse backgrounds can make it easier for someone in the community to spot such problems (Longino 1990, 197–94; Wylie 2003; Lacey 2005).

Nevertheless, we are all human, and there are no sure-fire ways to guarantee that we are not subtly presuming the very thing we wish to test. The best we can do is to acknowledge that values should not direct our choices in the early stages of science in such a pernicious way. As philosophers of science continue to push into this murky territory, some additional theoretical clarity on the relationship between research project selection and methodology formulation may be forthcoming. For now, let us note that allowing a role for direct values in these early stages of science should not be taken to include a direct role that undermines the value of science itself.

Even with this additional complexity on the table, it remains clear that some direct role in the early stages is acceptable, and that the full range of values (cognitive, ethical, and social) are relevant. Yet this direct role must be limited to the stages early in science, where one is deciding what to do and exactly how to do it. Once the study is under way, any direct role for values must be restricted to unusual circumstances when the scientist suddenly realizes that additional direct value considerations need to be addressed. For example, the scientist may come to understand that the chosen methodological approach, once thought acceptable, actually is ethically unacceptable, and that some modification is needed. This issue arises with medical studies that encounter unexpected adverse effects among test subjects. Such a reconsideration of methodologies, or of the stated purpose of a study, is the only acceptable direct role for values in science during the conduct of the study. Otherwise, values must be restricted to an indirect role.

Why would a direct role for values be consistently problematic in the later stages of science? Consider what would be acceptable reasoning if a direct role for values were allowed in the heart of doing science—during the characterization of data, the interpretation of evidence, and the acceptance of theories. A direct role for values in the characterization of data would allow scientists to reject data if they did not like it, if, for example, the data went against a favorite theory. The value the scientists placed in the theory could override the evidence. A direct role for values in the interpretation of evidence would allow values to have equal or more weight than the evidence itself, and scientists could select an interpretation of the evidence because they preferred it cognitively or socially, even if the evidence did not support such an interpretation. And if values were allowed to play a direct role in the acceptance or rejection of scientific theories, an unpalatable theory could be rejected regardless of the evidence supporting it. For example, it would be legitimate to reject a theory of inheritance that was politically unacceptable solely for *that* reason (as is the standard understanding of the Lysenko affair).[14] Or one could reject the Copernican theory of the solar system solely because it contravenes church doctrine, and ignore the evidence that supports the theory on the grounds that the theory is socially or ethically undesirable.[15] These problems are not just limited to social values in a direct role; they hold for cognitive values as well. We do not want scientists to reject theories solely because they are not simple, not as broad in scope as their competitors, or seem to have less explanatory power.

If we allowed values to play a direct role in these scientific decisions, we would be causing harm to the very value we place in science, the very reason we do science. We do science because we care about obtaining a more reliable, more true understanding of the world. If we allow values to play a direct role in these kinds of decisions in science, decisions about what we should believe about the world, about the nature of evidence and its implications, we undermine science's ability to tell us anything about the world. Instead, science would be merely reflecting our wishes, our blinders, and our desires. If we care about reliable knowledge, then values cannot play a direct role in decisions that arise once a study is under way. The potential for these disturbing possibilities underlies the worry that motivates support for the value-free ideal.

Despite these valid concerns about a direct role for values in science, an indirect role for values in science is still open at the heart of doing science. Such a role does not undermine the purpose of science. The value-free ideal is too strict, excluding needed ethical and social values from a legitimate indirect role, and thus preventing scientists from fulfilling their moral responsibilities to fully consider the consequences of error. The value-free ideal is also too lax, bringing in cognitive values in a way that is often too permissive, allowing them to potentially play a direct role. A better ideal is to restrict values of all kinds to an indirect role at the core of the scientific process.

Indirect Roles for Values in Science

The indirect role for values in science concerns the sufficiency of evidence, the weighing of uncertainty, and the consequences of error, rather than the evaluation of intended consequences or the choices themselves. Values should be restricted to this indirect role whenever the choice before a scientist concerns a decision about which empirical claims to make. If we value having a better understanding of the world, we do not want values *of any kind* determining the empirical claims we make. Our values, whether social, ethical, or cognitive, have no direct bearing on the way the world actually is at any given time.[16] Thus, when deciding upon which empirical claims to make on the basis of the available data or evidence, values should play only an indirect role.

Choices regarding which empirical claims to make arise at several points in a scientific study. From selecting standards for statistical significance in one's methodology, to choices in the characterization of evidence during a study, to the interpretation of that evidence at the end of the study, to the decision of whether to accept or reject a theory based on the evidence, a scientist decides which empirical claims to make about the world. At each of these decision points, the scientist may need to consider the consequences of error. If there is significant uncertainty and the consequences of error are clear, values are needed to decide whether the available evidence is sufficient to make the empirical claim. But note that values in the indirect role operate at the margins of scientific decisionmaking rather than front and center as with the direct role. Values weigh the importance of uncertainty, but not the claim itself. Evidence has a stronger role in determining the claims we make, with improved evidence reducing the uncertainty and thus the need for values. This should be the case for all kinds of values in science, whether cognitive, ethical, or social, when deciding upon empirical claims.

[margin note: ETHICAL]

First, let us consider the indirect role for values in methodological choices. As noted above, values can play a legitimate direct role in methodological choices, excluding ethically unacceptable methodological approaches. Care must be taken that values do not play a pernicious direct role, where research agenda and methodology serve to decide the outcome of research projects beforehand. In addition to these direct roles, there is also an indirect role for values in methodological choices. For example, the deliberate choice of a level of statistical significance is the choice of how much evidence one needs before deciding that a result is "significant," which usually is the standard to be met before a claim is taken seriously. This choice sets the bar for how much evidence one will demand, thus establishing the burden of proof. It requires that one consider which kinds of errors one is willing to tolerate.

For any given test, the scientist must find an appropriate balance between two types of error: false positives and false negatives. False positives occur when scientists accept an experimental hypothesis as true and it is not. False negatives occur when they reject an experimental hypothesis as false and it is not. Changing the level of statistical significance changes the balance between false positives and false negatives. If a scientist wishes to avoid more false negatives and is willing to accept more false positives, he or she should lower the standard for statistical significance. If, on the other hand, the scientist wishes to avoid false positives more, he or she should raise the standard for statistical significance. For any given experimental test, one cannot lower both types of error; one can only make trade-offs from one to the other. In order to reduce both types of error, one must devise a new, more accurate experimental test (such as increasing the population size examined or developing a new technique for collecting data).[17] While developing a more accurate experimental test is always desirable, it is not always feasible. It may be too difficult or expensive to expand a study population, or there may be no better way to collect data. Within the parameters of available resources and methods, some choice must be made, and that choice should weigh the costs of false positives versus false negatives.

Weighing those costs legitimately involves social, ethical, and cognitive values. For example, consider the setting of statistical significance levels for an epidemiological study examining the effects of an air pollutant on a population. A false positive result would mean that the pollutant is considered a health hazard when in fact it is not. Such a result would lead to unnecessary alarm about the pollutant and the potential for costly regulations that would help no one. In addition, scientists may mistakenly believe that the pollutant is dangerous at that exposure level, thus distorting future research endeavors. A false negative result would mean that the pollutant is considered safe at the measured exposure levels when in fact it is not. Such a result could lead to people being harmed by the pollutant, possibly killed, and scientists would more likely fail to follow up on the research further, having accepted the false negative result. The social and ethical costs are the costs of the alarm and the regulation on the one hand, and the human health damage and resulting effects on society on the other. The cognitive costs include the mistaken beliefs among scientists that shape their future research. How to weigh these costs against one another is a difficult task that I will not attempt here. However, some weighing *must* occur. All these kinds of values are relevant to the choice of statistical significance.[18]

Once the scientist has selected a particular methodological approach, including the level of statistical significance sought, the research project begins in earnest. In the collection and characterization of data, the scientist hopes that all goes as planned, and that no difficult cases arise requiring further judgment. However, this is not always the case. Often, scientists must decide whether an experimental run is a good one, whether to include it in their final analysis of the data or to throw it out. Robert Millikan, when performing his Nobel Prize–winning oil-drop experiment to measure the charge of an electron, had to decide which experimental runs were reliable and which were not. He discarded many runs of the experiment (Holton 1978). Sometimes scientists must also decide how to characterize an event or a sample. For example, if one is looking at a rat liver slide and it appears abnormal, is it in fact cancerous, and is the cancer benign or malignant? Experts in a specialized field can have divergent judgments of the same slides (Douglas 2000, 569–72). In cases where data characterization is uncertain, which errors should be more assiduously avoided?

Depending on the study, different kinds of values are relevant to weighing these errors. If one is looking at a toxicological study assessing the impact of a chemical on test animals, one must decide how to characterize tissue samples. Clear cases of disease should, of course, be characterized as such, as well should clear cases of healthy tissue. To not do so because one prefers not to see such a result is to allow values to play an unacceptable direct role at this stage, thus undermining scientific integrity and basic epistemic standards.

But there are often unclear cases, borderline cases where expert disagreement occurs and there is substantial uncertainty. How to characterize these cases? Eliminating them from the study reduces the sample size and thus the efficacy of the study. Characterizing the borderline cases as diseased to ensure that no diseased samples are missed is to worry about false negatives and the harmful effects of an overly sanguine view of the chemical under study. Characterizing borderline cases as not diseased ensures that no healthy tissue is mischaracterized as diseased, and will ensure a low number of false positives. This approach reflects worries about the harmful effects of having an overly alarmist view of the chemical under study. To split the difference between the two is to attempt to hold the value concerns on both sides as being equivalent.[19] The valuation of the consequences of error is relevant and necessary for making a decision in these uncertain cases. When one has judgment calls to make at this stage, values are needed to characterize data to be used in later analysis, although most scientists would prefer there be no borderline cases requiring a judgment call.

Finally, once the data has been collected and characterized, one needs to interpret the final results. Was the hypothesis supported? What do the results mean? Whether to accept or reject a theory on the basis of the available evidence is the central choice scientists face. Often, the set of evidence leaves some room for interpretation; there are competing theories and views, and thus some uncertainty in the choice. The scientist has a real decision to make. He or she must decide if there is sufficient evidence present to support the theory or hypothesis being examined.

As Rudner and others argued decades ago, the scientist should consider the consequences of error in making this choice. The implications of making a wrong choice depend on the context, on what the empirical claim is, and the implications of the claim in that context. In our society, science is authoritative and we can expect (even desire) officials to act on the basis of scientific claims. So the implications of error include those actions likely to be taken on the basis of the empirical claim. Such actions entail social consequences, many of which are ethically important to us. Therefore, social and ethical values are needed to weigh the consequences of error to help determine the importance of any uncertainty. Note that this role is indirect—it is only in the importance of the uncertainty that the values have any role to play.

Given the need for moral responsibility in accepting or rejecting empirical claims, it is clear that social and ethical values can and should play an important indirect role in these scientific choices. But what about cognitive values? Should we consider their role to be direct or indirect in these choices? It might seem at first that their role is direct, for do not scientists choose a particular interpretation of data because it has explanatory power?

However, simply because a theory has explanatory power is not a good reason on its own to accept that theory. Any "just-so" story has plenty of explanatory power, but we do not think Rudyard Kipling's explanatory tales, for example, are true or empirically reliable. Lots of theories have explanatory power but are still false, unreliable guides for understanding and making choices in the world. The explanatory power of Aristotle's geocentric universe did not make it true, nor did the explanatory power of phlogiston or ether save them from the dustbin of science.

Similarly, simplicity, scope, fruitfulness, and precision are also not reasons on their own to accept a theory. Occam's razor notwithstanding, a simple theory may not be a true or reliable one. A simple theory, though elegant, may just be wishful thinking in a complex world. A theory with broad scope may not be a true one. Diverse phenomena may not fall under the confines of one theory. A fruitful theory, leading to many new avenues of research, may prove itself false over time, even if it spurs on research. Even a theory that makes a precise prediction may not be a true one. Recall Kepler's model of the heavens as structured by platonic solids. It made rather precise predictions (for its time) about the spacing of the planets, but is now thought to be a quirky relic of the seventeenth century. None of these cognitive qualities increase the chance that a theory is reliable, that its acceptance is the right choice.[20] So what is the value of these cognitive values?

Despite the implausibility of a legitimate direct role for cognitive values, the presence of cognitive values aids scientists in thinking with a theory that exemplifies cognitive values. Thus, a theory with such values is easier to work with, to use, and to develop further. But the importance of this ease is not to make the scientists' work less demanding. Rather, it means that a theory reflecting cognitive values will be more likely to have its flaws uncovered sooner rather than later. Cognitive values are an insurance policy against mistakes. They provide a way to hedge one's bets in the long term, placing one's efforts in theories that, if erroneous, their apparent acceptability will not last long. Thus, scientists should care about cognitive values because of the uncertainties in scientific work, not because one really wants simpler, more broadly scoped, more fruitful theories for their own sake. If more evidence arises that reduces uncertainty in the choice, cognitive values become less important. A simpler or more explanatory theory with lots of evidence against it is not very attractive. A more complex theory with lots of evidence for it is attractive. (Hence, the demise of the simpler DNA master molecule theory in favor of a far more complex understanding of cellular interactions.) Cognitive values in themselves are no reason to accept or reject a theory, just as social and ethical values are no reason in themselves to accept or reject a theory. Cognitive values serve an indirect rather than a direct role in these choices, helping to mitigate the uncertainties over the long term.

Let us consider a general example to make this point clearer. Suppose one has a choice between (1) a theory with broader scope and explanatory power for a range of contexts and examples, and (2) a theory with narrower scope and similarly constrained explanatory power. The evidence for the narrower theory is slightly better than the evidence for the broader theory, but only slightly. Suppose we are working in a very theoretical area of science with no practical implications for error. A scientist would be wise to choose the broader scoped theory because with more areas for application one is more likely to find problems with the theory sooner rather than later. Similarly, a theory that makes precise predictions rather than vague ones will more likely expose errors sooner.

In sum, cognitive, ethical, and social values all have legitimate, indirect roles to play in the doing of science, and in the decisions about which empirical claims to make that arise when doing science. When these values play a direct role in the heart of science, problems arise as unacceptable reasoning occurs and the reason for valuing science is undermined. When values conflict in the indirect role, they must be weighed against each other. A brief account of one of the major medical mishaps of the twentieth century will help to illustrate this understanding of values in science.

Diethylstilbestrol (DES): Cognitive, Social, and Ethical Values in Practice

In 1938, a new compound was discovered that acted like estrogen but was much cheaper to produce and easier to administer. Diethylstilbestrol, or DES, was an exciting discovery in a period of intensive research into human hormones and how they influence human health. Researchers at the time believed that many "female problems" were due to a lack of sufficient female hormones, and thus the appearance of DES seemed a boon to those seeking to treat these female problems, which included menopause, mental health issues, venereal diseases, and miscarriages. By 1947, the FDA had approved DES for all of these medical uses.[21]

Yet there were reasons to believe that the use of DES, particularly in pregnant women, was risky. It was known that DES caused birth defects in animal studies, and it was known that it crossed the placental barrier. So why would scientists think using DES to prevent miscarriages would be a good idea? Some researchers had measured a drop in female hormones in the month before a problem arose in pregnancy. The prevalent view held that men and women were essentially biologically different and that hormones dictated much of this difference. If women experienced a drop in essential female hormones, it was little surprise to scientists of the time that the women would be unable to perform the essential female function of bearing children. The biological essentialism of the day, reflected in this theory of how hormones played a role in childbearing, was reinforced by the social norms of the day, that women and men should play very different social roles because of their biological differences. A decline in female hormones in a pregnant woman would understandably undermine the quintessential female function—childbearing. Thus, social norms and values reinforced the apparent explanatory power and simplicity of the overriding theory, that female hormone levels must be maintained at a high level for a successful pregnancy. Under this view, an estrogen imitator like DES would be naught but beneficial for any pregnant woman. By the early 1950s, tens of thousands of pregnant women in the United States were taking DES every year to assist with their pregnancies. The reasons and evidence to be concerned about DES were largely ignored.

In 1953, a careful controlled study was published which showed that DES produced no reduction in the number of miscarriages (Dieckmann et al., 1953). Despite this new evidence, many researchers refused to accept the study's implications and the FDA did not change its stance on DES. It just made sense to many scientists that DES was good for pregnant women. Earlier studies had lent some credence to the theory of DES's benefits for pregnant women, but none were as carefully done as the contravening 1953 study. However, there were theoretical reasons to reject the 1953 study. It was argued that the 1953 study did not have enough of a focus on high-risk pregnancies, even though the drug was being prescribed for many other pregnancies, and it went against the theory that female hormones were essentially good for female functions. In addition, no other drug was available at the time to help reduce miscarriages; bed rest was the main alternative. It was not until the early 1970s, when evidence emerged that DES could cause a rare and dangerous cancer in the female children of DES mothers, that DES was withdrawn for use for pregnant women.

In the face of substantial uncertainty, all three kinds of values played a role in helping scientists decide what to make of the available evidence in 1953. First, cognitive values reinforced belief in the beneficial efficacy of DES. If one believed that estrogen was the hormone that made everything work well for women, then more estrogen, even artificial estrogen like DES, would improve the womanliness of an expectant mother, thus increasing the health of the baby. Having a simple model of sex and hormones led some scientists to believe that DES was necessarily good for women.[22] In addition, the simple model of hormone function also seemed to have great explanatory power, allowing scientists to understand a range of gender differences. Thus, the theory that DES was good for pregnant women and their babies was supported by the cognitive values of simplicity, scope, explanatory power, and consistency with other theories of the day. Second, social values also seemed to support the use of DES. Beliefs about hormones bolstered faith in the social structure of the 1950s, and vice versa. If women's functioning was essentially determined by their hormones, then women were essentially different from men, and rigid gender roles were appropriate. In addition, if human gender roles were hormonally determined, then it seemed obvious that DES aided women in becoming even better women. The social value of stability in fixed gender roles lent credence to the idea that DES was good for women. Third, ethical values weighed in on both sides. If scientists pushed the use of DES and were wrong, women and their children would be harmed (as they in fact were). If scientists failed to push DES and it was beneficial, a valuable tool for pregnant women would be overlooked, and again women who wanted children and their miscarried fetuses would be harmed.

What does this episode tell us about these values in scientific reasoning? First, it illustrates nicely the way in which cognitive and social values can be mutually reinforcing. The social values of the day bolstered the apparent cognitive values of the theory in use, that female hormones are good for what ails women. The explanatory power of DES as good for women increased when one considered the socially prevalent gender roles, as the hormonal essentialism that explained the success of DES also explained the dichotomous gender roles found in society. And the apparent helpfulness of DES for pregnant women (based on anecdotal evidence or studies without controls) bolstered the sense of the essential gender differences pervasive in the social norms. As Rooney (1992) and Longino (1996) argue, the cognitive and the social can blur together.

Second, the case of DES demonstrates the dangers of using any values, cognitive or social, in a direct role for the acceptance or rejection of a hypothesis. The theory, that a lack in what makes a woman a woman (estrogen) was the cause of a problem in the essential function of womanhood (having children), had great explanatory power, scope, and simplicity. It also fit with the social norms of the day. Yet *neither* the cognitive values *nor* the social values are a good reason in themselves to accept the theory. Unfortunately, they seem to have played precisely such a direct role in this case. This is particularly apparent once the 1953 study appeared, and some key scientists preferred to continue to hold to their ideas about DES in the face of strong evidence to the contrary. Even if the cognitive values are divorced from social values, the explanatory power and simplicity of one's theories are still not good reasons to ignore evidence. The cognitive values competed with and trumped evidence in this case, as they should not.

Third, cognitive, ethical, and social values should play only an indirect role in the assessing of evidence and the choice of interpretation. Thus, the values should focus on the uncertainties in the case and help to weigh the sufficiency of evidence, not compete with the evidence. Ethical values should weigh the consequences of error in accepting or rejecting the theory underlying the use of DES to prevent miscarriages. Which is worse, to cause birth defects or to fail to use a method to prevent miscarriages? Given these potential consequences of error, what should the burdens of proof be for accepting or rejecting DES as a preventer of miscarriages? This is a difficult choice, and given such tensions, the scientist committed to pursuing DES as a treatment should (a) restrict the use of DES to high-risk cases only, and, in the meantime, (b) rely on cognitive values to hedge one's bets. The presence of cognitive values indicates which understanding of DES, if wrong, will be uncovered sooner, and should direct research efforts accordingly. Thus, the scientists should have used the noted simplicity and explanatory power to find flaws in the theory quickly; that is, they should have explored the implications of the underlying theory more rigorously, testing it for reliability in multiple contexts rather than relying on the theory *because* it was simple and had explanatory power. If this reasoning had been followed, either the use of DES by pregnant women would have been rejected (given the general weight of evidence by 1953 that DES did not help prevent miscarriages and that DES had the potential to cause substantial harm), or scientists who viewed the potential benefits as a strong enough reason to continue its use would have put more effort into determining the reliability of the underlying theory, and DES would have been pulled from the market long before the 1970s.

The scientists who supported DES use did not act in this way. Instead, the explanatory power, simplicity, and fit with social custom were taken as reasons for the correctness of the view. But values should not act as reasons in this manner; they should not be taken to provide warrant. It took decades for a more complex understanding of hormone functioning to emerge. The direct role for cognitive and social values hampered science and led to children being harmed by exposure to DES in the womb.

In sum, cognitive values should serve as heuristics to figure out how quickly and with what ease one might uncover an error. To think about cognitive values as indicators of reliability themselves is to make a serious error. Cognitive values themselves do not improve the likelihood that a choice is accurate or reliable. When present, they do indicate that one may discover a potential error more quickly than otherwise, *if* one pursues further research using the positive aids of the cognitive values to develop further tests of the theory.

Because cognitive values are no indicator of reliability, they should not function in the place of evidence, nor be on a par with evidence, just as social and ethical values should not function in that way. Both cognitive and ethical values function as ways to mitigate against uncertainty just in case one chooses incorrectly. Ethical values appropriately weigh the potential consequences of error, thus shifting burdens of proof to avoid the more egregious consequences. Cognitive values signal where error is more likely to be uncovered quickly by scientists, if error is present. Neither ethical nor cognitive values should play a direct role in deciding upon empirical claims. An indirect role, however, is pervasive and necessary, both to provide guidance when making judgments in the face of uncertainty and to meet the moral responsibilities of scientists.

Conclusion: Value-Laden versus Politicized Science

Science is a value-laden process. From the decision to do science, to the decision to pursue a particular project, to the choice of methods, to the characterization and interpretation of data, to the final results drawn from the research, values have a role to play throughout the scientific process. We need social, ethical, and cognitive values to help weigh the importance of uncertainty and to evaluate the consequences of error. In addition to this indirect role for values in science, there is also a direct role, crucial to some decisions in science. The direct role must be constrained to those choices in the early stages of science, where one is not yet deciding what to believe from one's research or what empirical claims to make. Once a scientist has embarked on the course of research, the indirect role for values should be the only role for values, whether social, ethical, or cognitive. To admit of a direct role of values in these later stages is to violate the good reasoning practices needed to obtain reliable knowledge about the world.

This distinction between the two roles for values in science allows for a clearer understanding of the concern over the politicization of science. In the cases of politicized science, the norm against a direct role for values in the decisions about empirical claims is violated. When scientists suppress, or are asked to suppress, research findings because the results are unpalatable or unwelcome, values are playing a direct role in the wrong place in science. When scientists alter, or are forced to alter, the interpretations of their results because they are unwelcome by their funders or their overseers, values again are playing an unacceptable, direct role.

In both of these kinds of cases, values are directing what we think, rather than helping us grapple with uncertainties. It is one thing to argue that a view is not sufficiently supported by the available evidence. It is quite another to suggest that the evidence be ignored because one does not like its implications. In the former case, we can argue about the quality of the evidence, how much should be enough, and where the burdens of proof should lie. These are debates we should be having about contentious, technically based issues. In the latter case, we can only stand aghast at the deliberate attempt to wish the world away. Allowing a direct role for values throughout science is a common way to politicize science, and to undermine the reason we value it at all: to provide us with reliable knowledge about the world.

The conceptual structure I have described in this chapter thus allows for a clear distinction between value-laden and politicized science. All science is value laden. But unacceptable, politicized science occurs when values are allowed to direct the empirical claims made by scientists. This understanding of values in science also addresses the concerns that underlie the value-free ideal. Values should never suppress evidence, or cause the outright rejection (or acceptance) of a view regardless of evidence. The classic cases of Galileo and Lysenko fall into unacceptable uses of values in this framework. By distinguishing between direct and indirect roles for values, the fundamental integrity of science is protected.

Science is saturated with values, and not only in the real and imperfect human implementation of scientific ideals. Values are needed in the indirect role to assess the sufficiency of evidence. There may be cases of scientific research where only cognitive values are relevant to these assessments, but these cases are merely simplified instances of the norms I have articulated (only one general kind of value is relevant rather than all kinds) rather than an ideal for which scientists should strive. Such "pure science" is the special case of science without clear ethical or social implications, one that can switch rapidly to the more general case when research scientists suddenly find themselves facing social implications. The new ideal I have articulated here should cover *all* of science. The importance of values in the heart of science decreases with decreasing uncertainty, but as long as science is inductively open, uncertainty is ineliminable, and thus so are values.

So science is saturated with values in the ideal—any ideal which claims for science a value-free zone in principle is an inadequate and incomplete ideal. This conclusion seems to leave us in a quandary. For so long, we have been convinced that the value-free nature (or ideal) of science was what made it valuable. Being value free gave science its objectivity; being value free gave science its superior ability to resolve disputes over contentious empirical issues. But the objectivity of science does not fall with the rejection of the value-free ideal. Value-saturated science can still be objective, as we will see in the next chapter.

CHAPTER 6

OBJECTIVITY
IN SCIENCE

THE VALUE-FREE IDEAL IS A BAD IDEAL for science. It is not restrictive enough on the proper role for cognitive values in science and it is too restrictive on the needed role for social and ethical values. The moral responsibility to consider the consequences of error requires the use of values, including social and ethical values, in scientific reasoning. Yet the inclusion of social and ethical values in scientific reasoning seems to threaten scientific objectivity. Our notion of objectivity should be reworked and clarified in light of the arguments of the previous two chapters. We need an understanding of objectivity that reflects its important role in our language, the complexity of our use of the term, and the moral responsibilities of scientists.

The complexity of usage arises because we call many different kinds of things "objective"—objective knowledge, objective methods, objective people, objective observations, and objective criteria, to name a few. As Lorraine Daston, a historian of the concept of objectivity notes, "We slide effortlessly from statements about the 'objective truth' of a scientific claim, to those about 'objective procedures' that guarantee a finding, to those about the 'objective manner' that qualifies a researcher" (Daston 1992, 597). Somehow, we know generally what we mean when applying the adjective "objective" in all these cases, but we do not mean precisely the same thing in each case (see also Daston and Gallison 1992, 82).

What holds all these aspects of objectivity together is the strong sense

115

of trust in what is called objective. To say a researcher, a procedure, or a finding is objective is to say that each of these things is trustworthy in a most potent form (see Fine 1998, 17–19). The trust is not just for oneself; one also thinks others should trust the objective entity too. Thus, when I call something objective, I am endorsing it for myself, *and* endorsing it for others. For example, when I call an observation "objective," I am saying that I trust the observation, and so should everyone else. Or if I state that a scientist is objective, I am saying I trust the scientist and so should everyone else (although what I am trusting the scientist for may be different than what I am trusting the observation for). Common to all the uses of objectivity is this sense of strong trust and persuasive endorsement, this claim of "I trust this, and you should too." It is this commonality that underlies the usage of objectivity in its various guises.

While this sense of trust and endorsement provides a common meaning for objectivity, the bases for trust vary with the different applications of the term. With objectivity applied to so many kinds of things, there are several bases for trust, several kinds of good reasons to believe something is reliable enough that others should trust it too. In this chapter, I will lay out seven bases for such a trust, to make clear how to understand the concept of objectivity in practice, and to elucidate the concept's fundamental complexity. I make no claims that the seven bases for objectivity I describe here are a *complete* account of the concept. Arguments could be made that there are additional bases for objectivity not discussed here. But the richness of the concept, and of its uses in practice, requires an account at least as complex as the one presented here.

My analysis of objectivity will focus on the objectivity of knowledge claims, and the processes that produce these claims, rather than the objectivity of persons, panels, or procedures per se. For my purposes, whether a claim can be considered objective, and thus trustworthy, is what we want to gauge. An objective person who makes no claims or an objective method that produces no claims is of little interest for the policy process. It is the trustworthiness of knowledge claims that is a central aspect of objectivity— often, objective people, methods, and measures are valued to the extent that they make knowledge claims that we care about, and on which we need to rely. In approaching objectivity in this way, we can focus on those aspects of the processes leading to knowledge claims that give us confidence in the trustworthiness of the claims. In short, for my discussion here, objective processes produce trustworthy knowledge claims. I will take the ascription of objectivity as a shorthand way of telling others that a claim is likely to be trustworthy, given the processes that produced it. Thus, I will analyze markers of objectivity in the processes that lead to knowledge claims. A different primary focus (on objective people or procedures) may produce a different topography of objectivity, although I would expect substantial overlap.

In describing objectivity in this way, it will become clear that <u>claiming something is objective is not an absolute statement</u>. For all of the bases for objectivity described below, <u>there are degrees of objectivity</u>, that is, there can be more or less of it. Thus, it makes sense to talk about a more or less objective process, person, or statement. <u>Objectivity is not an on or off property</u>. In addition, <u>none of the bases for objectivity should be thought of as a guarantee</u>. Even if some aspect of objectivity is robustly present, that does not ensure that one has it right, that the knowledge claim will be forever stable or acceptable. Objectivity can provide no such absolute assurance. Instead, <u>objectivity assures us that the best we can do is being done</u>. The more objective something is, the more we can trust it to be our best epistemic effort. Little more can be demanded, given the fallibility of actual human practice.

In the end, allowing values to play their proper role in science, "proper" as delineated in the previous chapter, is no threat to scientific objectivity. Once we have a deeper and more robust understanding of objectivity, we can see that scientists can (properly) use values in their reasoning and still be objective. Indeed, one of the senses of objectivity depends crucially on a proper role for values in scientific reasoning.

Processes for Objectivity

When I say that some knowledge claim is objective, I am endorsing that claim, stating that the claim can be trusted by both myself and others. <u>Objectivity does not ensure truth, but rather ensures the claim was the result of our best effort to accurately and reliably understand the world</u>. What do we look for before ascribing objectivity to a claim if we cannot assess the truth of that claim directly? We examine the process that leads to the claim to see if we have good reason to think that the outcome of that process is worth our endorsement. By examining the process that produced the claim, we will decide whether the claim deserves the attribute "objective."[1]

In general, three different kinds of processes deserve examination: (1) interactions that one can have with the world such as experimental processes, observations over time, or simple interactions in daily life; (2) an individual's apparent or reported thought processes, focusing in particular on reasoning processes that lead to certain claims; and (3) social processes, which are interactions or ways of doing things among groups of people. In each of these kinds of processes, we can find bases for trusting the outcomes of the processes; we find reasons to endorse the claims that come out of these processes. While I am most interested in objectivity in science here, there is much overlap between objectivity in science and objectivity in everyday existence. That some of these aspects of objectivity have manifestations in both scientific and nonscientific arenas should bolster our confidence that we are capturing the important aspects of objectivity.

Human Interactions with the World

When looking at human interactions with the world, what does it mean to say that such a process leads to something objective? For example, what does it mean to say that a particular experiment produced an objective result, solely in terms of the interaction between the human experimenters, their equipment, and their results? Two different processes produce knowledge claims that should be considered objective to some degree.[2]

The first process is one of manipulating the world, repeatedly and successfully. Suppose we have some experimental result, or empirical claim, which can then be used in multiple additional experimental contexts reliably. For example, consider the ability to manipulate the genetic material DNA to produce specific experimental animals, most notably mouse strains, to be used in further experiments. The claim that DNA is the genetic material of the animal is readily considered to be an objective claim here. When we can use a new concept or theory like a tool to intervene reliably in the world, ascriptions of the objectivity of those new tools come easily and with strong endorsement. When we can not only bump into the world in reliable and predictable ways, but also use the world to accomplish other interventions reliably and predictably, we do not doubt that we are using *something* and that it has some set of characteristics we are able to describe, and thus manipulate that something. As in Ian Hacking's example from *Representing and Intervening*, scientists do not doubt the objectivity of some claims about electrons when they can use them to produce images of other things with an electron scanning microscope (Hacking 1983, 263). I will call this sense of objectivity *manipulable* objectivity, in recognition of the process of manipulation or tool use central to its meaning.

This sense of objectivity is as important outside of the laboratory as inside. When we can use objects around us, we trust our accounts of their existence and properties as reliable. In other words, when we can reliably use the objects, the objectivity of the claims we make about those objects central to their use becomes apparent. If I can reach out and drink from a glass of water, and it quenches my thirst, and I can fill it back up again, repeating the whole process reliably, I have good reason to trust the reliability of relevant beliefs or claims about the glass, such as that it is in front of me, it is solid enough to hold water, it is light enough for me to pick up, and so forth. I will consider my beliefs about the glass to be objective, particularly those beliefs that enable me to get that drink of water. Note that this way of determining the objectivity of claims is probably just as important as having agreement among competent observers (a sense of objectivity which will be discussed below). In addition, this method of ascribing objectivity does not require more than one observer-participant, as long as that observer is able to repeat his or her interventions. When we can't repeat our interventions (if I can't pick up the glass again, if it dissolves under my grasp, or, in other contexts, if the experimental process fails to recur after repeated attempts), I rightly doubt the presence of manipulable objectivity, and thus the reliability of the results. Perhaps it was all an illusion after all.

Note, too, that even the strong sense of objectivity here, the ability to successfully intervene in the world, is not absolute. Depending on how successful and with what precision we can manipulate the world around us, the strength of the sense of objectivity will wax or wane. For example, the use of DES to prevent miscarriages (discussed in chapter 5) seemed to be a reliable intervention to those physicians prescribing the drug, but a more precise measurement of that intervention, in the form of a double-blind controlled study, belied the objectivity of that particular claim. Also, the particular claims we make about the objects may be more or less central to our success. While it may seem that the solidity of my water glass would indicate that the glass is not mostly empty space, further experimentation with matter, and an understanding of atomic theory, suggests that my claims about the nature of the solidity of the glass are not necessary to it functioning successfully in quenching my thirst. The solidity of the glass need merely rest on the strength of attraction among the glass molecules being sufficient to contain the water, and resist the pressure of my hand. The amount of empty space turns out to be irrelevant. In sum, we get degrees of manipulable objectivity by considering how reliably and with what precision we can intervene in the world, and how essential our claims about those interventions are to the success of those interventions.[3]

A second sense of objectivity is more passive than manipulable objectivity but may be nearly as convincing. Instead of using an initial result for further interventions, we approach the result through multiple avenues, and if the same result continues to appear, we have increasing confidence in the reliability of the result. I will call this sense of objectivity, *convergent objectivity*.[4] This sense of objectivity is commonly employed in both scientific research and everyday life. When evidence from disparate areas of research all point toward the same result, our confidence in the reliability of that result increases. Mary Jo Nye (1972) explicated this type of reasoning in the development of atomic theory. When multiple and diverse areas of research all pointed to the atomic nature of matter, the multiplicity and diversity of evidence convinced scientists of the objective nature of atomic theory, long before individual atoms were manipulable. In the work of Jean Perrin (1913, 215), in particular, one can see the building of an argument based on convergent objectivity. Perrin measured Avogadro's number (the number of molecules per mole) in several different ways, using, for example, the vertical density of an emulsion, horizontal diffusion in a liquid, and rotational Brownian motion in particles. Drawing upon the work of other scientists, Perrin detailed over a dozen different ways to calculate Avogadro's number, each producing the same (or nearly same) result. This convergence of evidence is what convinced scientists of the molecular nature of matter.

Cases in astronomy use similar approaches. When several information-gathering techniques provide different but related images of an object, one gains increasing confidence in the general result indicated (for example, that one has a certain kind of star or astronomical object in view with its associated properties). In everyday life, when an object continues to appear from a variety of vantage points and using a variety of techniques (for example, both sight and sound), the possibility of illusion seems remote.[5] As any birdwatcher will tell you, a convergence of evidence from various sources (such as bird coloration and song, along with habitat and time of year) assists greatly in the objective identification of the species under observation.

However, one must be aware of the limitations of convergent objectivity. The strength of the claims concerning the reliability of the result rests on the independence of the techniques used to approach it.[6] One must ask whether the techniques employed really are independent ways of gathering evidence or if there is some source of error (or a gremlin in the works) across the methods. Also, it is possible that the convergence is only apparent, and several distinct phenomena are combining to produce the results. One way to bolster the sense of convergent objectivity in the face of these concerns is to make predictions using the claim under scrutiny, particularly predictions about future events that would not be otherwise expected. Such predictions, if found accurate, allow for more supportive evidence for the claim. The more such predictions pan out, the more one has a sense of the independence of the claim from any particular context, or any lucky coincidence that might lead to a successful prediction, despite a flawed basis for that prediction. The more one can be assured of the independence of a claim from particular contexts or methods, the more convergent objectivity one has.

Both manipulable and convergent objectivity require examination of the experimental process (or more generally, human-world interactions) to find the process markers that support ascription of objectivity to (that is, endorsement for trust in) the results. For manipulable objectivity, we look for success in using the empirical claim to reliably and successfully intervene in the world. Empirical claims that allow us to intervene repeatedly and across multiple contexts are considered objective. For convergent objectivity, we look for multiple and independent (as possible) sources of evidence that point toward the same result or empirical claim. When we see this kind of diversity of evidence, we readily ascribe an objective status to the claim supported. With both convergent and manipulable objectivity, we have varying degrees of confidence that we have gotten objective results depending on our confidence in the independence of the methods and their specificity, or depending on the reliability of the results as a tool and the variability of contexts in which it is reliable. Thus, both senses of objectivity are matters of degree.

Individual Thought Processes

In examining interactions with the world, we focus on one kind of process central to objectivity. The thought processes of the individual producing the knowledge claims are also of central concern. When looking at individual thought processes, what does it mean to say that the end result of that process is objective, that the claims being produced are objective? For example, if we say that someone has written an objective overview of a problem, or produced an objective analysis of a situation, what do we mean? Instead of focusing on the interaction between the experimenter and the world, these aspects of objectivity focus on the nature of the thought process under scrutiny, and in particular on the role of values in the individual's thought processes. There are three ways in which objectivity of a thought process can be construed, only two of which are acceptable, in my view. The difficulty here is that the three are often conflated in practice, with deeply problematic results.

The least controversial and most crucial sense is the prohibition against using values in place of evidence. Simply because one wants something to be true does not make it so, and one's values should not blind one to the existence of unpleasant evidence. As Lisa Lloyd (1995) argues, "If one is personally invested in a particular belief or attached to a point of view, such inflexibilities could impede the free acquisition of knowledge and the correct representation of (independent) reality" (354). It is precisely for this reason that some metaphorical "distance" or detachment between the knower and their subject is recommended. Such detachment will keep one from wanting a particular outcome of inquiry too much, or from fearing another outcome to such an extent that one cannot see it. I call this sense *detached objectivity*.

This sense of objectivity is expressed in the prohibition for a direct role for values in the internal stages of scientific reasoning, the prohibition central to chapter 5. If values—social, ethical, or cognitive—take the place of evidence, or are considered good reasons to ignore evidence, the value we place in scientific knowledge is undermined. Acting against detached objectivity, allowing values to function in the same role as evidence in one's reasoning, damages the purpose of pursuing empirical knowledge, which is to gain knowledge about the world, not to gain an understanding that suits one's preferences. Values should not protect us from unpleasant or inconvenient evidence. Evidence needs to be able to challenge our current ideas and preconceptions, and be able to overturn them. Being open to challenge and revision in this way is part of what we value in science. Thus, detached objectivity is an essential part of our ideals about science.

Unfortunately, detached objectivity is often expanded to, and conflated with, *value-free* objectivity. In value-free objectivity, all values (or all subjective or "biasing" influences) are banned from the reasoning process. This potential aspect of objectivity derives support from the idea that values are inherently subjective things, and thus their role in a process contaminates it, making it unobjective. Only objective values could possibly be acceptable to such an ideal for objectivity, to assist scientists in making the judgments essential to science, yet which values could serve this role is unclear. As I argued in chapter 5, a clear demarcation between epistemic and nonepistemic values does not appear tenable, and even cognitive values raise disputes among scientists over their importance and interpretation. In addition, embracing this aspect of objectivity would require that we exempt scientists from their moral responsibility to consider potential consequences of error in their work, as such considerations require ethical values to weigh those consequences properly. Rather than attempt to uphold this aspect of objectivity, we should carefully examine whether it is in fact laudable (as I would argue it is not, based on the concerns raised in chapter 4), or whether it is instead a vestige of a postpositivist hangover that all values in the scientific reasoning process are bad, in which case we should discard it.

The effects of defining objectivity as value-free are problematic for the practice of science and our proper evaluation of scientific work. We train scientists to believe that values are not allowed in science and they must ward off any appearance of values or judgments playing a role in their doing of science. As a result, scientific papers have a very formulaic structure in which the role of the scientist as active decisionmaker in the scientific process is deftly hidden. Yet hiding the role of values and judgment in science is a mistake. Hiding the decisions that scientists make, and the important role values should play in those decisions, does not exclude values. It merely masks them, making them unexaminable by others. The difference between detached objectivity and value-free objectivity is thus a crucial one. It is irrational to simply ignore evidence, but it is not irrational, for example, to consider some errors more grievous than others (and thus to be more assiduously avoided) or to choose a particular avenue of investigation because of one's interests. Scientists need to acknowledge the important role values must play in scientific reasoning, while not allowing values to supplant good reasoning. The vigilance scientists have so long directed toward keeping values out of science needs to be redirected toward keeping values from directly supplanting evidence and toward openly acknowledging value judgments that are needed to do science. In other words, scientists must learn to negotiate the fine but important line between allowing values to damage one's reasoning (for example, blotting out important evidence or focusing only on desired evidence) and using values to appropriately make important decisions (such as weighing the importance of uncertainties). It is that distinction that defines the first sense of objectivity (detached) and is obliterated by the second (value-free).

There is a third sense of objectivity that is also often conflated with value-free objectivity but plays an important and distinct role in modern practice. This sense, *value-neutral*, should not be taken to mean free from all value influence. Instead of complete freedom from values, this sense focuses on scientists taking a position that is balanced or neutral with respect to a spectrum of values. In situations where values play important roles in making judgments but there is no clearly "better" value position, taking a value-neutral position allows one to make the necessary judgments without taking a controversial value position and without committing oneself to values that may ignore other important aspects of a problem or that are more extreme than they are supportable. It is in this sense that we often call a written overview of current literature "objective."[7] It takes no sides, makes no commitments to any one value position, but it adopts a "balanced" position. While the overview may in fact incorporate values in how it presents and views its topic, it does not allow extremes in those values.

Thus, with value-neutral objectivity, "objective" can mean reflectively centrist. One needs to be aware of the range of possible values at play, aware of the arguments for various sides, and to take a reflectively balanced position. Such value-neutrality is not ideal in all contexts. Sometimes a value-neutral position is unacceptable. For example, if racist or sexist values are at one end of the value continuum, value-neutrality would not be a good idea. We have good moral reasons for not accepting racist or sexist values, and thus other values should not be balanced against them. But many value conflicts reflect ongoing and legitimate debates. Such conflicts arise, for example, in debates between those placing primary value on robust local economies based on industrial jobs and those placing primary value on preventing health and environmental harms potentially caused by those industries. Another example would be the conflict between the needs of current generations around the globe, and the potential needs of future generations. In these and similar cases, value-neutral objectivity would be a clear asset.

If we reject value-free objectivity as an undesirable ideal, we are left with detached objectivity and value-neutral objectivity as capturing the relevant desirable characteristics of individual thought processes. I have suggested that detached objectivity is crucial for doing science, and that it reflects the norms for values in science articulated in chapter 5. Value-neutral objectivity is also useful for some contexts. Both senses of objectivity provide bases for trust in the outcome of the process; both provide reasons to endorse the claims that arise when they are manifest. And both can be manifest in degrees. One can be more or less detached from one's subject and thus more or less successful at keeping personal values from directly interfering with one's reasoning. One can be more or less neutral with respect to various values, more or less reflective on the spectrum of values and positioned in the middle of extremes. We can determine the degree of objectivity by examining the reasoning process and looking at the role of values, or, for value-neutrality, by ensuring that one has considered the range of values and has taken a middle position considerate of that range. Examining the thought processes of individuals and their interactions with the world does not exhaust the processes relevant for objectivity, however.

Social Processes

In addition to the two kinds of processes discussed above, we can also examine social processes when looking for bases for objectivity. When examining the social processes involved in knowledge production, what does it mean to claim that the end result is objective? Instead of examining an individual line of thought, one examines the process used among groups of people working to develop knowledge, and specifically, the process used to reach an agreement about a knowledge claim. There are three distinct senses of objectivity concerned with social processes, each with different epistemological strengths and weaknesses.

Social processes can be considered "objective" if the same outcome is always produced, regardless of who is performing the process. I will borrow from Megill (1994) and label this sense *procedural* objectivity.[8] Procedural objectivity allows society to impose uniformity on processes, allowing for individual interchangeability and excluding individual idiosyncrasies or judgments from processes. If there is a very clear and rigid quantitative form with which to process information, regardless of who processes that information, the same outcome should result. For example, in the grading of procedurally objective tests, no individual judgment is required to determine whether an answer is correct or not (such as multiple choice tests, once the key is made). The tests are designed so that there is one definitively correct answer for each question. Thus, there is no need for individual judgment on the correctness of a given answer. It is either correct or not, and the final score is the sum of correct answers. Regardless of who grades the tests, the same score should result for each test. If the same score does not result, one looks for errors in the grading, not points of judgment where disagreement may occur.

Theodore Porter's historical work traces the development of this sense of objectivity in the past two centuries (Porter 1992, 1995).[9] In his examination of objectivity in accounting, he shows how the focus on rules, particularly inflexible and evenhanded ones, lent credence to the field of accounting (1992, 635–36). In *Trust in Numbers* (1995), Porter expands the areas of examination, looking at the role of rule-bound quantification techniques across engineering, accounting, and other bureaucratic functions. Quantification through rules (as opposed to expert judgment) allows for both an extension of power across traditional boundaries and a basis for trust in those with power. Procedural objectivity thus serves a crucial function in the management of modern public life.

The key to procedural objectivity is that regardless of who engages in a procedurally objective process, if they commit no clear errors, they do it in the same way, producing the same result. While enabling public trust, the elimination of personal judgment and thus idiosyncrasies does not ensure the elimination of values. Instead of an individual's values playing a role through their own judgments, values are encoded in the processes themselves. A rigid quantitative process that eliminates the need for personal judgment forces one to examine the situation in a fairly narrow way. Inflexible rules mean that some nuances, which might be important to particular cases, will be left out. As Porter (1992) states, "Quantification is a powerful agency of standardization because it imposes some order on hazy thinking, but this depends on the license it provides to leave out much of what is difficult or obscure" (645). Which inputs are emphasized as important for the decisionmaking process reflects whatever values are built into the process. Thus, rules can force one to disregard evidence that one might otherwise consider relevant. The way that outcomes are determined in the process can also reflect values. Such processes generally have set thresholds for when one outcome as opposed to another will be warranted. Whether those are the appropriate thresholds for the outcomes reflects a value judgment, one made when the procedure was initially structured. Thus, procedural objectivity, while eliminating individual judgment, in no way eliminates values. Fine (1998) notes, "Bias and the impersonal are quite happy companions" (14).

Procedural objectivity is not the only way we look to social processes for an objective outcome. Two additional senses of objectivity are often subsumed under the heading of "intersubjectivity." The first, which I will call *concordant* objectivity, reflects a similar concern with unanimity of outcomes across groups of people. Instead of seeking to eliminate individual judgment, however, this sense checks to see whether the individual judgments of people in fact do agree. When we say that an observation is objective if some set of competent observers all concur on the particular observation, it is this sense which we are using. While procedural objectivity may be useful for achieving concordant objectivity, it is not necessary for it.[10] The processes leading up to the agreement may be socially sanctioned and fairly rigid, forcing individual judgment from the arena, or they may be looser, requiring judgment from individuals. A context that relies on individuals using their own judgment to come to an agreement without procedural constraints would provide a stronger sense of concordant objectivity than a context with procedural constraints. Presumably there would be greater potential sources for disagreement without procedural constraints. Thus, to reach agreement without them would increase the apparent reliability of the group's judgment.

For Quine (1992), this sense of objectivity is essential: "The requirement of intersubjectivity is what makes science objective" (5). As Quine and others have used it, concordant objectivity is applied in cases where the individuals are simply polled to see how they would describe a situation or context, or whether they would agree with a particular description. There is no discussion or debate here, no interactive discourse that might bring about agreement. If the agreement is not there, there is no concordant objectivity. If the observers agree, then the observation is concordantly objective. This aspect of objectivity is either present or it is not, *given any particular context*. The aspect gains degrees of more or less objectivity by looking at the details of the context—for example, whether a context is more or less procedurally constrained, or whether there are more or fewer observers present. It is in these details that one can assess the strength or weakness of concordant objectivity.

While concordance among a set of observers can be a powerful statement, the limitations of concordant objectivity must be recognized. Concordant objectivity cannot guarantee that one grasped something real; there is always the chance of a group illusion. It cannot guarantee that values are not influencing or supplanting reasoning; the observers may have shared values that cause them to overemphasize or to disregard aspects of an event. Nevertheless, *idiosyncratic* values that influence observation will not be allowed to determine a statement of fact. Idiosyncrasies may prevent concordance, but only widely shared idiosyncrasies (an oxymoron) can help create it. If agreement is attained, the testimony of the group will bolster claims for the actuality of the observation. There is a second limitation: how one decides on the composition of the group, a weakness shared with the third sense of objectivity.

This third sense of objectivity, *interactive* objectivity, involves a more complex process than concordant objectivity. Instead of simple agreement, this sense of objectivity requires discussion among the participants. Instead of immediately assenting to an observation account, the participants are required to argue with each other, to ferret out the sources of their disagreements. It is in the spirit of this sense that we require that scientific data be shared, theories discussed, models be open to examination, and, if possible, experiments replicated. The open community of discussion has long been considered crucial for science. The hope is that by keeping scientific discourse open to scrutiny, the most idiosyncratic biases and blinders can be eliminated. In this way, other people help to make sure you are not seeing something just because you want to.

The social character of science and its relationship to objectivity has been an increasingly important topic among philosophers of science (Kitcher 1993, Longino 1990, Hull 1988). As Longino argues, "The objectivity of science is secured by the social character of inquiry" (1990, 62). She goes on to discuss conditions of social interaction that increase the interactive objectivity of a process of inquiry. Among these are recognized avenues for criticism, shared standards for arguments on which criticisms are based, general responsiveness to criticism, and equal distribution of intellectual authority (76–79). The quality of interaction among investigators and the conditions for those interactions are crucial for interactive objectivity.

Among the difficult details of interactive objectivity, including precisely how the discussion is to be structured and how it is to proceed, one of the most difficult issues remains defining who gets to participate in the discussion. Having a set of communal standards for discussion requires that some boundaries be set up between those who agree to the shared standards and those who do not. In addition, some discussions require a degree of competence or skill, so for both concordant and interactive objectivity, defining competence becomes a problematic issue. Some level of competence is essential (minimally that there are shared language skills), but requiring very high and uniform standards for competence reduces participant diversity. Some diversity among participants is crucial for interactive objectivity; getting agreement among a group of very like-minded people increases the possibility that a shared bias or delusion will produce an unwarranted result. All the observers in agreement may share the same problematic bias, particularly in a low-diversity group, and this bias would probably go unnoticed. Thus, a deep tension remains under both interactive and concordant objectivity—between shared standards that provide a basis for discussion and agreement and the diversity of participants. Because of this tension, neither interactive nor concordant objectivity can guarantee that all ideological bias has been removed from science.

This tension and the other complications in applying interactive objectivity allow for degrees of objectivity. According to Longino 1990, "A method of inquiry is objective to the degree that it permits *transformative* criticism" (76). Transformative criticism comes about from diverse perspectives producing critiques that are heard and taken seriously. Depending on how strong the requirements are for the diversity of the group sitting at the discussion table, how the process of discussion is structured (for example, is intellectual authority equally distributed among participants?), and how stringent the requirements are for agreement (what is going to count as consensus?), the end result will be more or less objective. Similarly for concordant objectivity, the greater the diversity of participants, the greater the degree of objectivity. (The other aspects of practices central to interactive objectivity are not important for concordant objectivity, as it assumes equality in intellectual authority among its participants and there is no discussion.) As is the case with all seven acceptable sense of objectivity for knowledge claims, objectivity is a matter of degree.

Synergism among the Bases for Objectivity

One might be overwhelmed by the complexity of objectivity at this point. I have described seven normatively acceptable but different bases for objectivity above (manipulable, convergent, detached, value-neutral, procedural, concordant, interactive), drawing from three different kinds of processes (human-world interactions, individual thought processes, social processes). I have argued elsewhere that these seven aspects of objectivity are conceptually distinct and cannot be reduced to each other (Douglas 2004). Even though all seven acceptable bases for objectivity are distinct, they can also act in concert to strengthen one's sense of objectivity in a certain claim. Because each basis for objectivity should act to bolster our confidence in the reliability of the knowledge claim being made, multiple bases acting at once increase the strength of our endorsement of the claim. Two examples drawn from current environmental issues and their associated scientific work illustrate this point.

Consider first the observations in 2004 and 2005 of an ivory-billed woodpecker in the swamps of Arkansas. Biologists had not observed the bird in the United States since 1944, and the last reliable reports of the species outside the United States were from the 1980s in Cuba. Many assumed the bird to be extinct and considered the loss of America's largest woodpecker one of the great environmental tragedies of the twentieth century. But in February 2004, two naturalists believed they spotted an ivory-billed woodpecker in Arkansas's Cache River National Wildlife Refuge. It was the first sighting of the bird in the United States by two people, both thought to be reliable observers, in sixty years.[11] The report of the sighting drew ornithologists from Cornell University to the site to attempt to gather more evidence. At least five additional sightings by bird experts were made in the following year, including a rather grainy video of the bird in flight. The ornithologists argue that they saw distinctive patterns, found only on ivory-billed woodpeckers, and that they can also be seen in the video (Fitzpatrick et al. 2005). In addition, sound recordings of the birds' distinctive calls were made (Chariff et al. 2005).

The scientists held off announcing this remarkable find for a year, not releasing any reports until April 2005 to help ensure that their claim to finding an ivory-billed woodpecker was as objective as possible. In doing so, they drew on several aspects of objectivity discussed above. First, they used different kinds of information to confirm the presence of the bird. In particular, they used both visual information (the bird's distinctive flight pattern and wing feathers) and audio information (the bird's calls and distinctive raps). In addition, the habitat was perfect for the bird. These sources provided a sense of convergent objectivity. Second, the researchers worked hard to maintain detached objectivity, trying not to get too excited about the possibility of the bird's survival in the United States. To help maintain their detachment, they planned carefully structured expeditions into the area where the bird would be expected, and they systematically took appropriate data. This helped to maintain the detached objectivity needed. Finally, they brought in additional researchers to see if they too could spot the bird. This was not to be a matter of discussion, but more a matter of whether different people would simply see the same thing during a bird sighting. Here, the researchers were seeking concordant objectivity. In sum, convergent, detached, and concordant objectivity were all part of making the claim, that ivory-billed woodpeckers were still surviving in the United States, objective.[12] Unfortunately, additional attempts to find the bird have been less successful (as of this writing). Perhaps the earlier sightings were of the last of its kind, and it is now extinct. Perhaps the researchers were mistaken, despite the reasons we have for thinking the initial claim to be objective. This example serves to emphasize that objectivity, even when present in multiple senses, does not guarantee a particular result. It does, however, assure us that we are making our best epistemic efforts.

A more complex example of synergistic interactions among the senses of objectivity can be found in work on climate change. Attempting to measure changes in the earth's climate is admittedly a daunting task, given how large and variable the planet is. Yet there has arisen an increasing sense of objectivity about the claim that the earth's climate has warmed substantially in the past decade. (The cause of this warming is another, even more complex topic.) Several aspects of objectivity are at play here. First, the ways in which to measure the earth's current and past climate have grown substantially.

Not only do we have land-based temperature records and sea temperature records from the past century, we now have satellite data for the past two decades.[13] Initially it was thought that the satellite readings were diverging from the older form of measurement. However, systemic errors were found in the satellite readings, and once corrected, the two forms of measurement agreed (Santer et al. 2003; Kerr 2004). In addition, temperature assessments from glacier movements, reef behavior, sea level measurements, and species movements and behavior are now being recorded. All of these various sources indicate the same strong warming trend from 1995–2005. Such convergent objectivity provides one basis for trust in the assessment of recent climate.

The reasoning of climate scientists and their group processes also provide bases for trust. In their work, climate scientists take pains to remain detached, trying not to see warming that is not there or to miss warming that is. Detached objectivity is crucial in such a complex field, where it is difficult to have clear predictive tests of models. Social interactions also help bolster trust. Climate scientists from around the world have met repeatedly since 1988 through the auspices of the Intergovernmental Panel on Climate Change (IPCC). Thousands of scientists from more than 150 countries have participated in these meetings, representing a wide range of social and ethical perspectives on the subject. With such a large sample of scientists, the IPCC could be considered value neutral. Perhaps most importantly, many conferences and discussions have been convened to interpret the various sources of information about the climate. These conferences, structured well for full and open discussion, help to produce interactive objectivity.[14]

As these two examples illustrate, different aspects of objectivity may be more or less important to the assessment of the emerging claims, depending on the context. When large groups of scientists are needed to assess or produce a claim, value-neutral or interactive objectivity can become central. When smaller numbers of scientists are involved, concordant objectivity comes to the fore. Detached objectivity is always an asset and can be assisted by concordant or interactive objectivity—it is difficult for scientists to continue to insist on a claim for which they have no evidence but only a preference in the face of disagreement and critique by their fellow scientists. Procedural objectivity is useful when one has a clearly defined and accepted process that assists with knowledge production (such as a standard and straightforward lab technique). Both convergent and manipulable objectivity provide a way to directly assess the evidence available for the claim. And if one has concordant or manipulable objectivity, interactive objectivity, and a clear consensus among scientists, the claim is more likely to obtain. The aspects of objectivity work together to provide bases for trust.

While one can expect some synergism among these bases, one should not then think that perhaps there is some fundamental basis for objectivity on which all the others rest. None of these bases is plausibly reducible to the others (Douglas 2004); objectivity is an inherently complex concept. Being explicit about when and where we should ascribe objectivity to a claim, about the bases for that ascription, will help us to better understand the nature of the persuasive endorsement behind the ascription of objectivity. It also allows objectivity to be a practically understandable and applicable concept rather than a merely rhetorical term of endorsement.

Conclusion

This understanding of objectivity for knowledge claims is a rich one. If we understand objectivity as indicating a shared basis for trust in a claim, we have multiple bases from which to draw in our ascription of objectivity to the claim. It is this richness that allows us to ascribe objectivity to a number of different kinds of claims arising from different processes, and for the concept to be useful across many contexts. Being clear about these bases enables us to ascribe objectivity to claims more precisely, and with clarity of reasons for the ascription. The complexity of objectivity is not a weakness, but a strength, for the aspects of objectivity can work together to reinforce our sense of the trustworthiness of a claim.

This complexity also allows for an understanding of what objective science would look like without the value-free ideal. I have described seven different bases for objectivity, none of which depend on science being value-free. Rejecting the ideal of value-free science need not be a threat to objectivity in science. We can understand and assess the objectivity of specific scientific claims even as we accept the proper role for values in science.

CHAPTER 7

THE INTEGRITY OF SCIENCE IN THE POLICY PROCESS

THUS FAR, I HAVE ARGUED THAT SCIENTISTS, when making judgments in their work, have a moral responsibility to consider the consequences of error, including social and ethical consequences, a responsibility that cannot be readily shifted to other parties. If they have this responsibility, then the proper role for values in science is not captured by the value-free ideal, and a new ideal is needed. This ideal must protect the integrity of science while allowing scientists to meet their general responsibilities. I have presented such an ideal, centered on the practice of limiting all kinds of values to an indirect role when making judgments concerning the acceptability of data and theories. With this ideal for values in science, we can still maintain a robust understanding of objectivity, with multiple aspects available for assessing the objectivity of a claim.

While these philosophical arguments may seem far from the practical realm of science in public policy, they hold some significant insights for how to understand the role of science in policymaking. In chapter 2, I described the rise of the scientific advisor in the United States. By the 1970s science advising had become a fully institutionalized aspect of the policy process. Although the place of the science advisor had become secure, there was an intensification of debate over which science was sufficiently reliable for policy decisions. The reality of dueling experts undermined the idea that more science in policy would make the policy process more rational and thus less contentious. Instead, science seemed to make the process more contentious and protracted, adding another focal point over which opposing interests could fight.

As the disputes became more public and apparently more intractable, there were attempts to impose order on the process of using science to inform policy by creating a more procedural approach to the use of science. These efforts were aimed at carefully defining the place of science in the policymaking process, with an eye toward addressing three concerns: (1) to generate uniformity in the judgments needed to assess the import of the available science, (2) to protect the integrity of science from politicization, and (3) to ensure democratic accountability in policymaking. These issues are serious concerns, and ones that the value-free ideal seemed to help meet. However, a more nuanced understanding of scientific judgment, including the pervasive need for social and ethical values, requires a change in the norms for science in the policy process. Adherence to the value-free ideal has structured our understanding of science in the policy process, including what it means to protect the integrity of science in that process. Science with integrity is usually considered science that has maintained complete disentanglement from social or ethical values. This, I will suggest, is a mistake.

We must tread carefully, however, and not neglect important science-policy considerations in addition to concern over scientific integrity. In particular, there are the concerns over maintaining the proper locus for decisionmaking authority in the policy realm. These concerns arise because the ultimate source of political authority for any democratic government is the citizenry. In the United States, the agencies that generate regulations do so by virtue of legislative authority from Congress, and are accountable to Congress for their regulatory decisions. Congress in turn represents the people of the United States, who hold their representatives to Congress to account on election day for their legislative actions (at least in the ideal). The agencies that write regulatory rules do so to implement the laws crafted by Congress. Thus, the federal agencies are legally responsible for final decisions ("rule-makings") at the regulatory level, and agencies are held accountable by Congress through public hearings and budget appropriations (Fiorino 1995, 67–70). Because of the need for a line of accountability, final regulatory decisions cannot be made by scientists brought in to advise the agency; usually they are made by politically appointed officials within the agency that can then be held accountable by elected officials. The holding of decisionmaking authority by the appointed official means that the scientific input can be only advice.

Yet it is extremely difficult for agency officials to ignore such advice. Science has an important *epistemic* authority in society. When questions about the workings of the natural world arise, science is generally the most reliable source to which we can turn for answers, and thus what scientists have to say on topics within their areas of expertise carries with it a certain cultural weight. Ignoring such advice can be a major source of embarrassment for a government agency and can ultimately undermine the *political* authority of the agency. The need for expert advice, as agencies deal with questions about new substances or processes, about which only particular scientists have any knowledge, makes it difficult to put aside such advice once given. This is particularly true after the passage of the Federal Advisory Committee Act (FACA), which requires that such advice be made public.

In the midst of this complex terrain, one must also keep in mind the difficulties inherent in providing helpful science advice. To be effective advisors, scientists must provide clear advice relevant to the policy process. Because of the complexity of available evidence pertinent to most policy cases, scientists, when assessing the evidence in any given case, must make judgments about the quality of available evidence, about what to make of conflicting evidence, and how a set of evidence from various sources should be assessed as a whole. If scientists are to provide useful advice to policymakers, they *must* make such judgments to make any sense of the available data. These judgments include assessments of the sufficiency of evidence to warrant a claim. Scientists who neglect important uncertainties in order to influence the final regulatory decision attempt to usurp the authority of the agency. It is the agency's responsibility to reflect the will of the people (as reflected by Congress), and to determine the appropriate value trade-offs (guided by the relevant laws) inherent in any regulatory decision.

In light of these complexities, it is not surprising that the value-free ideal for science would be attractive for science in policymaking. One method to ensure that scientists are not usurping democratic authority is to demand that no social or ethical values "infect" the science used to make policy, so that the science has no political valence generated by these values. In other words, one attempts to keep science "value-free" in precisely the sense defined by philosophers in chapter 3, a science insulated and isolated from the relevant social and ethical implications. This has been the standard norm for the role of science in policy for several decades, as reflected in Peter Huber's definition of sound science versus junk science, and as reflected in calls for the separation of science from policy (Huber 1991). For example, in 2002, Linda Rosenstock and Lore Jackson Lee mounted a spirited defense of evidence-based policy against pernicious forces that "are committed to a predetermined outcome independent of the evidence" (14). Such forces would clearly be in violation of the requirement that values be kept out of a direct role when assessing evidence, and that in determining empirical claims, values not compete with evidence. The authors identify several methods used by these pernicious forces to undermine sound science, from delaying tactics to harassment of scientists who attempt to publish undesirable findings, and Rosenstock and Lee attempt to find practical remedies for the problems. However, the upshot of these discussions is couched in the standard way, that in order to prevent attempts to undermine sound science, we must "more rigorously define and clarify the boundary between science and policy" (17). The authors assume that there is an obvious boundary to be policed, and that keeping science separated from the policy realm is essential for both science and policy. If this were desirable, it would indeed assist the proper functioning of democracy. However, because of the need for social and ethical values (often at the core of political debates) in the assessment of the significance of scientific uncertainty, such a boundary is bound to fail.

New norms are needed for science in policymaking, norms that center on clarity and explicitness about the bases for judgments that scientists must make, including social and ethical values in their proper roles, instead of on the maintenance of a clear science-policy boundary and the reinforcement of the value-free ideal. As I have emphasized in previous chapters, scientists should be considering the social and ethical implications of error in their work, even if this brings values, and thus some political valence, to the heart of their work. Rather than attempting to purify the science, we should clarify the nature of the judgments needed in science so that the responsible decisionmakers can be more fully informed about the nature of the science advice they are receiving, and thus make appropriate and accountable decisions on the basis of that advice. Concerns over the integrity of science are addressed not by excluding values, but by ensuring that values play only normatively acceptable roles in science, such as an indirect role only when assessing evidence. Scientists also must make the values in the indirect role as explicit as possible to assist the policymaker.

The first part of this chapter describes the traditional view of science in policymaking as an enterprise to be protected from influence by social and ethical values, and explains why this view fails to achieve its goals. This view was clearly articulated in the early 1980s and codified in the distinction between risk assessment and risk management. Eliminating the need for social and ethical values in risk assessment has been a challenge, however. I will examine this challenge in some detail, looking at the struggles over the inference guidelines that were constructed to help manage the judgments made by scientists in risk assessments, to make them more transparent, consistent, and accountable. Despite attempts to formalize inference guidelines, which have been accompanied by much controversy, the need for case-by-case judgment has not been eliminated. Because case-specific judgments are needed endemically in risk assessment, and because of the responsibilities of scientists, the desired purity of risk assessments cannot be achieved.

In the second part of this chapter, I will elaborate on what the failure of the traditional view means for the integrity of science in policymaking. Spelling out more precisely what we should mean by integrity for science helps make sense of what we should mean by sound science versus junk science as well. Maintaining scientific integrity (and avoiding junk science) becomes a minimum threshold that must be met; other aspects of objectivity from chapter 6 provide additional assurances for sound science above the minimum. Finally, I will begin to address what this altered view of science in policymaking means for the practices of scientists and of policymakers. Democratic accountability is assisted rather than hampered by the proper inclusion of values in science for policy, even if it makes the relevant government documents more cumbersome and policymaking more complex.

Managing Judgment in Science for Policy

In the early 1970s in the United States, the major concerns over science in policy centered on the openness of the advising process, spurred by the controversies that brought an end to the Presidential Science Advisory Committee in the Nixon administration and that led to the passage of FACA. By the mid-1970s, these concerns had broadened to include general democratic considerations of whether and to what extent the public should have input into technical regulatory decisions, and whether public openness would hamper the effectiveness of science advice. For example, the National Research Council's 1975 report *Decision Making for Regulating Chemicals in the Environment* noted that, in general, "A major concern about an open decision-making process is that it will excessively politicize the process and thus minimize the role of scientific expertise" (NAS 1975, 24). However, the panel disagreed with this sentiment, instead arguing "that openness and scientific input tend to be complementary rather than contradictory" due to the need to gain outside (of government) scientific advice and to air scientific considerations before an engaged public (24). As the controversies heightened over science in policy during the 1970s, the need for openness was overshadowed by the need to identify and effectively utilize reliable scientific input.

The issue of obtaining needed scientific advice became particularly prominent with the passage of new legislation that strengthened federal regulatory control over many areas, increasing the need for technically based policymaking. The Clean Air Act, the Federal Water Pollution Control Act, the Toxic Substances Control Act, the Resource Conservation and Recovery Act, the Safe Drinking Water Act, the Occupational Safety and Health Act, and the Consumer Product Safety Act all were passed in the 1970s.[1] These laws required newly created agencies (EPA, OSHA, and CPSC) to protect the public. But for each law, different standards of adequate protection, as well as different kinds of substances and routes of exposure, were to be considered by the agencies. For example, the Clean Air Act required the EPA to protect the public from air pollutants with "an ample margin of safety," but the Toxic Substances Control Act required that the EPA protect the public from "unreasonable risk" due to exposure from toxic chemicals (quoted in NRC 1983, 44–47). As agencies struggled to implement the new laws, and needed scientific information about a wide range of substances and exposure routes, the agencies' use of science to assess hazards became a focus of controversy.

Some had hoped that the need for more science to justify regulations would help policymaking become a more rational process, but by the late 1970s ongoing disputes about the science in many cases dashed such hopes. In 1979, Judge David Bazelon wrote, "We are no longer content to delegate the assessment and response to risk to so-called disinterested scientists. Indeed, the very concept of objectivity embodied in the word disinterested is now discredited" (277).

The intensity of the disputes over science in policy was leading to the idea that scientists could not be expected to be disinterested participants. It seemed that politics was infecting science, making it more political, and thus more contentious. If the usefulness of science for policy was to be redeemed, science needed greater protection from politics. To accomplish this, a clearer role for science in policymaking was proposed to ensure that apolitical science could properly have its say. A new understanding of science in policymaking was to help insulate science from politics as much as possible while integrating it into the policy process. The model was couched within the new framework of *risk analysis*.

This emphasis on risk analysis was instigated by a shift in the 1970s away from a focus on absolute safety in regulatory efforts to one centered on risk. The shift occurred for two reasons: (1) the technical ability to detect potentially dangerous chemicals at very low concentrations sharpened, and (2) increased concern over cancer, birth defects, and other harms more subtle than acute toxic poisoning brought into question the widely held assumption that a clear threshold always exists for the ability of a substance to harm humans. When even minute quantities of chemicals can be measured in environmental samples, and those quantities may have a chance, albeit a small one, of causing harm, then it becomes very difficult to ensure complete safety. Instead of absolute safety, one must accept that there are chances for harm (that is, risk) and then one must decide which risks are significant and which are insignificant. With absolute safety no longer a viable goal, the importance of good measurements of risk increased, so that risk could be managed to an acceptable level, be it a level "with an ample margin of safety" or with a "reasonable risk."[2]

The new focus on risk, the improved ability to detect and measure small quantities of chemicals, and an increased concern over long-term health risks pushed the limits of what could be inferred from the available evidence. This exacerbated the problem of uncertainty in science and increased the potential points of contention over science in policy. It seemed that agencies were forced to rely upon what Alvin Weinberg labeled "trans-science." Weinberg suggested that some apparently scientific questions were not yet answered definitively by science, and that this realm of questions should be considered "trans-science" (Weinberg 1972). Yet as the controversies continued, it became apparent that almost any uncertainty could be seized upon by opponents of regulatory actions, and thus become a driver of controversy.[3] In the regulatory realm, what might once have been thought of as settled science could *become* trans-science, as uncertainties took on new importance. Because the science of assessing the hazards posed by new chemicals in a range of environmental or workplace contexts was just beginning to be developed, a substantial level of uncertainty surrounded many assessments. This uncertainty, and the judgments agencies were forced to make in the face of that uncertainty, further fueled the controversies.

By 1980, the need for some management of science in policy was apparent. The newly formed arena of risk analysis became the space in which discussions over the role of science in policymaking developed. Risk analysis was divided into two distinct parts: risk assessment, where science could provide insight on the extent and nature of a risk, and risk management, where decisions could be made about how to handle the risk in practice. Conceptually dividing risk analysis into risk assessment and risk management served several purposes, but primarily it attempted to keep certain values as far from scientific judgments as possible, in honor of the value-free ideal. However, it is at these points of judgment that an indirect role for values is needed, given the often heightened uncertainty in the science and the important implications of making an incorrect judgment. Because of this problem, the attempts from the 1980s were not successful at dispelling the problem of dueling experts, as neither could the need for judgment be eliminated nor could a full range of value considerations be properly kept out of risk assessments.

The origins of the risk assessment/risk management distinction can be traced back to the 1970s. William Lowrance (1976) articulated a distinction that was to become central for the view of how regulatory agencies should handle risk regulation.[4] Lowrance divided the regulatory process into two phases: (1) measuring risk, and (2) judging safety, or judging the acceptability of the risks measured and what actions should be taken (Lowrance 1976, 75–76). Measuring, or "assessing," risk required four steps: "1. Define the conditions of exposure; 2. Identify the adverse effects; 3. Relate exposure with effect; 4. Estimate overall risk" (18). Once completed, the overall estimated risk was to become "the principal input to the heavily value-laden public tasks of judging the acceptability of risks and setting standards, regulating the market, planning research, and otherwise managing the hazard as a public problem" (41). The estimated risk "is the output of the scientific effort" and "becomes the input to personal and social decision making" (70). Thus, by separating "measuring/assessing risk" from "judging safety," Lowrance intended to separate the scientific component of the process—measuring the risk—from the value-laden social and political component of the process—deciding what to do about the risk. As Lowrance states, measuring risk "is an empirical, scientific activity" and judging safety "is a normative, political activity" (75–76). He claims that confusing the two is often a source of public turmoil: "Although the difference between the two would seem obvious, it is all too often forgotten, ignored, or obscured. This failing is often the cause of the disputes that hit the front pages" (76).

Lowrance's distinction between an insulated scientific component and a contentious, value-laden, political component became fundamental to the conception of risk analysis: risk assessment is the determination of what the risks are, whereas risk management involves decisions about what actions to take in light of the risks. In 1983, the NRC defined the two in the following way: "Risk assessment is the use of the factual base to define the health effects of exposure of individuals or populations to hazardous materials and situations. Risk management is the process of weighing policy alternatives and selecting the most appropriate regulatory action, integrating the results of risk assessment with engineering data and with social, economic, and political concerns to reach a decision" (3).

Risk assessment is to provide a technically based, scientifically sound assessment of the relevant risks, and is thus to be in the realm of science, the realm of fact as opposed to value. Risk management is to consider options to deal with that risk, including the possibility that the risk might be deemed acceptable. The contentious political wrangling is to be contained in the risk management part of the process, and kept out of the risk assessment.[5] In this model, science is to inform our understanding of the risks and causal factors at play, and *then* the messy ethical values complicating policy enter the fray.

There are several advantages to such a conceptual separation, particularly if it is then implemented as a procedural separation, which completes the risk assessment prior to embarking on a risk management.[6] First, it does seem to provide some insulation for the science-based part of the process, the empirical assessment of the risks. If the risk assessment is completed prior to the risk management, the risk assessment cannot be molded to fit emerging political agendas arising from the risk management process. Risk assessments *should* be protected from pressures to shift the assessment of risk because the results are politically inconvenient (a violation of the prohibition against the use of direct values in these kinds of judgments). Second, it was hoped that risk assessment would provide a clear place for an efficient summary of the available science, thus keeping scientists from having to get entangled in risk management.

Lowrance was quite right that the deeply political process of deciding which risks are acceptable and what to do about them should not be confused with assessing the nature and extent of risks. On the one hand, the acceptability of risk is an ethical, social, and ultimately a democratic issue, rather than a scientific one. Simply because a risk is assessed does not mean we must act to mitigate it. On the other hand, we do not want to succumb to the temptation of wishing risks away that might lead to uncomfortable political problems. Despite the apparent importance of the basic conceptual distinction, a closer look at the practice of risk assessment reveals the difficulty of maintaining the absoluteness of the distinction.

Because of uncertainty in the relevant science and the need for judgment to make use of the science, risk assessments cannot be completely insulated from policy concerns. Even while positing the conceptual distinction in 1983, the NRC recognized that because of incomplete scientific information, some inferences would have to be made in order to bridge the gap between generally accepted scientific information and an assessment of risk that would be of use to policymakers (NRC 1983, 48). Without such inferences, scientists are often unable to say anything specific enough to be of help to policymakers. In order to complete a risk assessment, scientists drafting (or reviewing) risk assessments must decide which inferences should be made in the face of uncertainty. This decision necessarily involves judgment. For example, in considering chemical risks to human health, one often needs to extrapolate from high experimental doses given to lab animals to low environmental doses to which humans are exposed. There are a range of plausible extrapolation methods available, including linear extrapolation and threshold models, which can produce a wide range of final risk estimates (several orders of magnitude apart) from the risk assessment. Rarely is there scientific information available that could justify a clear preference for one model over another. Qualitative uncertainty is thus endemic in risk assessment (given the current state of science), and inferences based on judgment are needed to bridge the gaps. In the same way that scientists must make judgments concerning uncertainty in their own work, risk assessors must make such judgments when completing a risk assessment.

In general, the NRC recognized that such inferences "inevitably draw on both scientific and policy considerations" (NRC 1983, 48). For example, scientific considerations, such as the currently accepted pathways for producing cancer, and policy considerations, such as how protective of public health an agency is required to be by law, should shape these inferences. However, the NRC did not think that the policy considerations *specific to the case at hand* should have an influence in shaping the inferences risk assessors made.

In order to shield agency scientists from the political pressures that arose in risk management, the NRC recommended that agencies develop "inference guidelines," defined as "an explicit statement of a predetermined choice among options that arise in inferring human risk from data that are not fully adequate or not drawn directly from human experience" (ibid., 51). Thus, instead of a choice shaped by the particulars of the specific policy case, the choice should be predetermined by general policy and scientific considerations. General policy considerations could include the nature of the law to be implemented and the degree of protectiveness desired by Congress or the relevant agency. Scientific considerations could include general knowledge about animal models or the understanding of biochemical interactions. The inference guidelines, determined by policy ahead of time, would ensure (1) that science was insulated from political pressures to manipulate judgments to produce a desired outcome, and (2) that the agency's official decisionmakers, who would approve these guidelines, would be accountable for the judgments made.

An example will make the nature of inference guidelines clearer. Regulatory agencies generally wish to be protective of public health and thus will adopt moderately "conservative" inference guidelines. ("Conservative" in this context denotes a tendency to prefer overestimates to underestimates of risk, in the name of protecting public health.) It is rare that one has definitive evidence that a substance will cause cancer in humans, because direct laboratory testing of humans is morally unacceptable and epidemiological evidence is usually inconclusive, unable to definitively isolate a single cause from the complex background of real life. If a substance was known to cause cancer in laboratory animals but there was no clear data for humans, one would have to make an inference from the available animal data in order to state anything clear concerning the risk posed to humans by that substance. One standard inference guideline is to assume that if a substance causes cancer in laboratory rodents, it will likely cause cancer in humans and should be considered a likely human carcinogen. Given that most mammals have similar responses to many substances, such an assumption has some scientific basis, but the blanket assumption of similarity also has a policy basis of being protective of human health. Once one makes this assumption of similarity among mammals, rodent data (derived from high doses) can then be used to extrapolate risks of cancer to humans (usually exposed to low doses) and to set acceptable exposure levels. This method of extrapolation involves yet another inference guideline and additional concerns over the degree of protectiveness.

The key, according to the 1983 NRC committee, is to make these decisions ahead of time, in the form of explicit inference guidelines, so that the risk assessment process, while not completely scientific, is not unduly influenced by the *specific* economic and social considerations of the case at hand, and thus can retain its integrity. As the NRC saw it, the maintenance of such integrity was central to the reasons for a distinction between risk assessment and risk management: "If such case-specific considerations as a substance's economic importance, which are appropriate to risk management, influence the judgments made in the risk assessment process, the integrity of the risk assessment process will be seriously undermined" (NRC 1983, 49). In other words, one reason for making a distinction between the two phases was to defend the integrity of the first phase, risk assessment, so that science could be protected from political pressures. Properly developed inference guidelines could maintain this boundary between the risk assessment and risk management—the latter containing case-specific values, the former being free of them (ibid., 69).

Inference guidelines had other benefits as well. In their overview of inference guidelines across agencies, the NRC noted that having guidelines made the risk assessment process more transparent, and thus more predictable. If the inferences to be made in the face of scientific uncertainty are explicit, interested parties can both anticipate and understand better the final outcomes of agency risk assessments. In addition, the NRC suggested that such guidelines could improve consistency in risk assessments, both within a single agency and, if guidelines were coordinated across the government, among agencies dealing with similar cases (ibid., 70–72). Such consistency assists with the appearance of fairness as well as with the general predictability of the outcomes given an available data set. The NRC committee also hoped that inference guidelines would help increase the efficiency with which scientific information could be made relevant to policymaking. With agreed-upon inference guidelines in place, the policy decisions that underlay the guidelines would not have to be revisited each time they were used (ibid., 73–74). All of these potential benefits rested on the hope that inference guidelines could reduce—or even eliminate—the need for individual judgment in the face of scientific uncertainty for each specific case, thus bringing more consistency, transparency, efficiency, and predictability to the use of science in policymaking.[7]

Despite these benefits, the NRC recognized potential problems with inference guidelines. Most important was (and remains) the problem of rigidity versus flexibility (ibid., 78–79). The more rigid the guidelines, the more predictable and transparent their use would be, and the more consistently they would be applied (ibid., 81). Such rigidity helps eliminate the need for judgment and the values that come with judgment, whether cognitive, social, or ethical. However, such rigid guidelines can run roughshod over the scientific complexities of specific cases. Consider the example of the inference guideline discussed above, that risk assessors should infer human carcinogenicity from rodent carcinogenicity. There are cases where the cancer does appear in chemically exposed laboratory animals, but we have good reason to think that cancer will not appear in similarly exposed humans. For example, exposure to some chemicals causes the livers of male rats to produce excessive amounts of a protein known as alpha-2u-globulin. When this protein accumulates in the rat's renal system, it harms the kidney and can induce cancer as a result of the tissue damage (Neumann and Olin 1995). Human (and female rat) livers do not produce this protein (Hard 1995). Chemicals that induce cancer only through this mechanism (and appear only in the kidneys of male rats) are not likely to be carcinogens in humans.

With a fuller understanding of the mechanism (and a fairly simple mechanism at that), the inference from rat cancer to human cancer appears unwarranted. An inference guideline that required the inference from rat cancer to human cancer would ignore important scientific evidence; an inference guideline that considered the additional evidence would require a case-specific judgment on the weight of evidence needed to override the default inference guideline, thus undermining the key advantages of the guideline (consistency, transparency, efficiency, and predictability) as well as exposing the risk assessment to questions of whether it was completed with integrity. The very openness of science to new and surprising evidence, which can challenge accepted theory, makes the acceptance of rigid inference guidelines problematic.

This difficulty with inference guidelines has not been resolved.[8] Recent EPA attempts to revise their cancer risk guidelines have struggled with precisely this issue. In their 2005 *Guidelines for Carcinogen Risk Assessment*, the EPA decided largely in favor of a high level of flexibility in the guidelines, requiring much individual judgment, even though this undermines the central goals of inference guidelines. Nevertheless, the agency still wishes to utilize inference guidelines in order to retain both rigidity (and the benefits thus produced) and flexibility: "The cancer guidelines encourage both consistency in the procedures that support scientific components of Agency decision making and flexibility to allow incorporation of innovations and contemporaneous scientific concepts" (EPA 2005, 1-2). After noting the need to balance these two goals, the EPA emphasizes two general ways in which the agency can have the flexibility needed to accommodate new scientific developments. First, they can issue supplemental guidance for particular topics, which may potentially diverge from previously accepted guidelines. Second, the "cancer guidelines are intended as guidance only," not as statements of binding regulation (ibid., 1-2). Thus, the EPA is not legally bound to follow the guidelines.

Within this general framework, the specific guidelines are also flexible, as illustrated in the language of default assumptions. Consider this sample default assumption: "In general, while effects seen at the highest dose tested are assumed to be appropriate for assessment, it is necessary that the experimental conditions be scrutinized" (ibid., A-3). This gives latitude to the risk assessor to accept or reject experimental findings based on the specific conditions of the testing (for example, whether or not doses are high enough to cause liver cancer in a testing animal purely through excessive cell death—caused by acute toxicity—and replacement proliferation, suggesting that the substance is not *really* a carcinogen). As the EPA notes, whether or not cancer found in the animal is a direct result of the substance being tested or a side effect of general toxicity in the animal being tested at high doses "is a matter of expert judgment" (A-4). Individual judgment will be needed to interpret the evidence and the experimental conditions that produced it, within the application of the inference guidelines.

The need for judgment is made even more explicit in the "weight of evidence narrative" that shapes the direction of the risk assessment and assigns a "descriptor" to the substance under scrutiny (ibid., 1-11). This narrative is produced in "a single integrative step after assessing all the lines of evidence" and "provides a conclusion with regard to human carcinogenic potential" (1-11–1-12). The descriptors to be chosen include such categories as "carcinogenic to humans," "likely to be carcinogenic to humans," "suggestive evidence of carcinogenic potential," "inadequate information to assess carcinogenic potential," and "not likely to be carcinogenic to humans" (2-54–2-58). These descriptors are to be justified by the weight of evidence narrative, which is to "lay out the complexity of information that is essential to understanding the hazard and its dependence on the quality, quantity, and type(s) of data available" (2-50).

Although the document provides some guidance on how this complexity is to be ultimately weighed, the EPA is quite clear that "choosing a descriptor is a matter of judgment and cannot be reduced to a formula" (ibid., 2-51). The goal of flexibility in relying upon the weight of evidence approach is affirmed: "These descriptors and narratives are intended to permit sufficient flexibility to accommodate new scientific understanding and new testing methods as they are developed and accepted by the scientific community and the public" (2-51). Although the acceptance of case-specific judgment allows the scientist to include the most recent and complex science in his or her risk assessment, it also works against one of the original purposes of the inference guidelines: to reduce the need for individual judgment. The more individual judgment involved, the less the process will be efficient, predictable, transparent, and consistent.

Even with these moves toward greater flexibility and greater need for individual judgment, the EPA still embraces the concern over integrity expressed in the NRC's 1983 report, which argued that case-specific social and economic considerations be excluded from risk assessment. As noted in the EPA's 2005 cancer guidelines, "Risk assessments may be used to support decisions, but in order to maintain their integrity as decision-making tools, they are not influenced by consideration of social or economic consequences of regulatory action" (ibid., 1-5–1-6). The prohibition on social or economic case-specific considerations is still upheld by the EPA today, twenty years later.

The endemic need for judgment in risk assessments, even with well-developed inference guidelines, poses some difficulties for the ideal of integrity for risk assessments espoused by the NRC and the EPA. Without the inclusion of case-specific economic and social considerations, the judgments, particularly those assessing the sufficiency of evidence for a claim, would be severely skewed. The EPA states that it upholds a policy of being protective of public health (because of its legislative mandate), but general protectiveness alone cannot adequately guide such judgments. If the agency were only following its statutory mandate to protect public health, all risk assessments would be excessively bleak, written as alarming as scientifically plausible in every case. Including the scientific considerations would not adequately protect against this outcome, as any significant uncertainty in the science could be amplified if the only allowable social consideration is the protection of public health. For example, any evidence of carcinogenicity that is scientifically plausible, regardless of concerns about the specific experimental conditions, should be taken as at least suggesting the substance is a likely carcinogen. Only if one takes into account some concern, even minimal, about overregulation, does the balancing seen in actual risk assessments become possible.

In addition, eliminating the case-specific considerations prevents the scientists who are making the judgments in the risk assessment from properly considering the consequences of error for that judgment. The more subtle *case-specific* considerations of both the potential threat to public health presented by a substance (is this chemical widespread and thus many people might be exposed?) or the ease of regulation (is this chemical easy to replace in industrial processes?) are needed in risk assessments if the moral responsibilities of chapter 4 are to be met. Because the judgments to be made are about assessing data and the support of evidence for a claim, the values that weigh these case-specific considerations should play an indirect role only. Risk assessments should not be determined by such considerations, and a direct role for social and economic case-specific considerations should not be tolerated. Nevertheless, judgments needed to complete risk assessments, including the application of inference guidelines, often involve judgments of whether the evidence is sufficient in that case. Is the available evidence sufficient for either the selection of a particular descriptor or the departure from a particular default assumption? As I discussed in chapter 5, assessments of the sufficiency of evidence involve the indirect use of values. The needed values include the social and economic consequences of both underregulation and overregulation, and the extent of our concern about both, for the case in hand.

If this is the case, then the traditional view of science in policy—that science must be kept as isolated as possible from policymaking in order to prevent its infection by values—should be discarded. In risk assessment, the judgments needed in the translation of science to policy cannot be eliminated. Nor can they be managed in a uniform manner, as the ever-present need for flexibility in inference guidelines demonstrates. Because of the endemic need for judgment, scientists performing those judgments cannot escape their moral responsibilities to consider the consequences of error when making these judgments. Weighing such considerations requires the use of case-specific values, including social, economic, and ethical. Yet the integrity of science in policymaking is still very important. What should we require for scientific integrity, given the complexity of the science in policy terrain?

Integrity in Science for Policy: Sound versus Junk Science Revisited

My argument for including case-specific social and ethical considerations in risk assessment may seem to strike at the integrity of science to be used in policymaking, particularly as it has been understood over the past two decades. But if we utilize the distinction between direct and indirect roles for values in scientific reasoning, integrity can be more carefully defined and precisely understood, capturing the main concerns about political influence on science. In addition, this refined account of integrity can illuminate the sound science–junk science debate, with junk science clearly delineated as science that fails to meet the minimum standards for integrity. Moving above and beyond those minimum standards enhances our confidence in the reliability of the science. Finally, requiring scientists to be open and up-front about the weighing of evidence, and the considerations that go into their judgments, can assist with ensuring the proper democratic accountability for regulatory decisions.

Although the NRC and the EPA have considered the inclusion of case-specific social and ethical values inimical to the integrity of science, I argue in chapter 5 that it is a *direct role* for values in making judgments about the acceptability of empirical statements that is inimical to science's integrity. The direct role for values that places them on a par with evidence would allow scientists to ignore evidence because they prefer a different outcome for a risk assessment. Rather than eliminate altogether social and ethical values from science and risk assessment, risk assessors should ensure that the values be constrained to the indirect role. Only in the face of uncertainty, and concerns over the sufficiency of evidence for a claim, can values properly enter into scientific reasoning. There, they act to determine whether the risks of error in making a claim are acceptable or not. Integrity should consist of keeping values in an indirect role only in risk assessment, in maintaining *detached* objectivity, not *value-free* objectivity.

With this understanding of how social and ethical values can play a legitimate role in science-based risk assessments, it is more understandable how experts could disagree about the extent of risks posed by a substance. The dueling experts of the past few decades do not have to misunderstand each other's subdisciplinary expertise, nor do they need to be dishonest practitioners, selling out the integrity of their expertise. Instead, they can (and often do) have very different views on the acceptability of uncertainty in the same case, even when looking at the same evidence.

An epidemiologist looking at a data set, for example, may worry more over the possible confounders that muddle a causal connection, particularly if they worry about the effects of overregulation on a company or society as a whole. The uncertainty created by confounders increases in significance because of a social value. A different epidemiologist looking at the same data set may worry about using the general population as a control for a group of workers, which can mask excessive disease among the workers. Because workers are generally healthier than the population as whole, workers compared statistically to the population as a whole should appear healthier. Thus, being merely as healthy as the general population may be a sign of excess disease in the group of workers. This epidemiologist may be worrying more about missing possible disease among the workers than in proclaiming a substance to be dangerous when it is not (for other examples, see

Douglas 1998, chap. 2). The significance and weight of uncertainty leads them evaluate the data differently. Dueling experts arise.

Frustration over dueling experts has fueled the sound science–junk science debates and the efforts described in chapter 1 to keep junk science out of the regulatory process. We must be careful about what we call "junk science," however. Merely the presence of disagreement among experts does not mean that someone is putting forth junk science. Instead, experts may be seeing different potential errors as more worrisome, and thus may interpret the available evidence differently, albeit within the bounds of acceptable scientific practice.

There are, of course, cases of junk science. These arise from three different types of missteps, one or more of which can be present in any given case. First, some scientists speak on issues outside their area of expertise, misunderstanding the work of their colleagues. Such examples can be deeply misleading to the public, who are unaware of the specialties of scientists. Second, there are incompetent scientists, unable to do acceptable work within their supposed area of expertise. Their poor methodology or unsupported inferences can lead to junk science. Only peer review (which should keep such work from being published) and public announcements pointing out the errors of such work when it gains a forum can remedy such junk science. Finally, there are those who do science without integrity, who are willing to say anything to support a predetermined outcome, regardless of the available evidence, or who are willing to set up a methodology that will guarantee a particular outcome to their liking. Scientists who cherry-pick evidence from a broad evidential record to support their views produce this kind of junk science, and this tendency has had a pernicious effect in science policy debates. For example, in the global warming debate, continued reference to the uneven rise in temperature in the twentieth century (a warming, cooling, then warming period) without the attendant complexity of aerosol emissions (which cool the atmosphere and were particularly high during the cooling period) is a classic example of junk science created by scientists ignoring relevant and widely known available evidence (Douglas 2006). Producers of such junk science undermine the general authority of science and must be called to account by the scientific community. But mere disagreement need not indicate the presence of any of these three sources of junk science.

So how does one detect junk science if mere disagreement is not enough to warrant the charge? The three different sources of junk science described above call for different methods, with varying levels of difficulty for the observers and users of science. The first kind of junk science—that produced by experts speaking outside their area of expertise—is relatively easy to detect. One needs merely to discover the nature of the expert speaking, which requires only a quick Internet search to find out in what area the expert has a Ph.D., in what area the expert has conducted research and published findings, and in what area he or she is employed (for example, the nature of the academic department or laboratory in which the expert works). If none of these seems to be relevant to the point on which the expert is speaking, one has good reason to suspect he or she may be speaking outside of the needed expertise, and thus potentially producing junk science.

For the second source of junk science—incompetent experts—a bit more work is required to uncover the problem. Such experts may well have the appropriate training, and may be employed in an apparently reputable setting, but their work is rejected as sloppy or methodologically flawed, and thus their findings are unreliable. For a person outside of the area of expertise to be able to detect this kind of junk science requires a bit more legwork. First, one should make sure the work is published in a journal with a peer review process and a robust editorial board. Are scholars from good universities on the board? Are they in the relevant disciplines? And are authors required to submit their work for peer review prior to publishing? These things can be determined by examining the Web sites for the journals, including the author submission guidelines. Finally, if the work has already been published in a reputable journal, are there retractions of the work in later issues? If the work is severely flawed and thus junk science, a good journal will note the flaws, even if they are uncovered only after publication.

The third source of junk science, which arises from a lack of scientific integrity, is the most problematic, both because it probably undermines science the most and because it can be the most time-consuming to uncover. Unless one is familiar with the evidential terrain, it can be difficult to know that a scientist is only focusing on that which supports his or her views, to the exclusion of well-known counterevidence. In order to reveal this kind of junk science, one needs to (1) become familiar with the scope and range of available evidence so that it is clear when evidence is being ignored; and/or (2) follow an open scientific debate for a while to see if the participants take into account the criticisms of their work and alter their views accordingly.

One of the key markers of junk science arising from a lack of scientific integrity is an unwillingness to acknowledge criticism or to moderate one's view to accommodate criticism. Not all criticisms may be on target, but steady and pointed critiques must be addressed. Thus, if one scientist raises an apparently relevant critique, the other should respond in some way. This takes time to play out, and the outsider to the debate must follow the back and forth to discover if junk science is present. The requirement that scientists take evidence seriously, and not ignore it even if it is problematic for their views, can be uncovered in another way as well. If an expert cannot even *imagine* evidence that could cause the expert to change his or her views, he or she is lacking in the requisite intellectual integrity. If no evidence imaginable could overturn the scientific claim, it is no longer a claim based on empirical evidence, but rather one based solely on the value commitments of the scientist.

With this understanding of junk science, sound science can be defined as science done competently and with integrity. One point must be clear, though: sound science need not produce consensus among scientists. Even scientists capable of doing sound science can legitimately disagree about how to interpret a body of evidence. Consider, for example, a recent evaluation of a draft risk assessment of dioxin conducted by the EPA's Science Advisory Board (SAB). In 2000, the SAB, the EPA's premier peer review panel for its science-based documents, was asked to review a draft of the EPA's dioxin risk reassessment.[9]

Among the specific questions on which the SAB was asked to focus was the issue of whether the EPA's designation of dioxin as a human carcinogen was sufficiently supported. The SAB panel split on the issue (EPA/SAB 2001, 13, 4). Some members of the panel thought there was sufficient evidence for the designation; others did not. The strength and weaknesses of epidemiological evidence, the relationship between that evidence and the effects seen in laboratory animals, and how the current understanding of the biochemical mechanisms of dioxin affect the understanding of the other evidence were all debated. Disagreement sprung not from a lack of understanding of the evidence, nor from a lack of integrity among participants, but from disagreement over the strength of the available evidence, and whether it was sufficient to overcome panel member doubts (ibid., 44–46). Divergent concerns over the consequences of error drive disagreements over the sufficiency of evidence. However, because under the value-free ideal scientists are prohibited from discussing such value considerations, the source of disagreement and of divergent expert opinion remains obscured.

If scientists are not able to express fully the bases of disagreements, the values that drive the disputes remain unexpressed, and thus unexamined by experts, government officials, and the public. Rather than rest integrity on the elimination of such values from the risk assessment process, the integrity of risk assessment would be strengthened if values could be clearly expressed as a legitimate basis for certain judgments. If scientists could state why they find a body of evidence sufficient for a claim, including concern over the consequences of error, both policy officials and the general public could more readily determine whether values are playing an acceptable indirect role (for example, "my concerns over public health lead me to find this body of evidence sufficient for seeing this substance as a human carcinogen"), and not an unacceptable direct role (for example, "because of my concerns over public health, no evidence could convince me that this substance is not a danger to the public").

Indeed, given my account of junk science, we should be asking scientists, particularly those with outlier positions, what evidence could conceivably convince them of a specific claim about which they are skeptical. If no possible evidence could convince an expert of an empirical claim, we can surmise that values, rather than evidence, are the primary support for their rejection of the claim, an improper direct role for values and an indicator of junk science. Being clearer about the assessment of uncertainty and its significance will assist with maintaining the integrity of science and the reasoning process employed when science informs policy, by allowing others to see whether values are playing a direct or indirect role.

If scientists could be explicit about values used in reasoning, the integrity of science would be maintained and the advising process would become more readily accountable and thus democratic. With the values used by scientists to assess the sufficiency of the evidence made explicit, both policymakers and the public could assess those judgments, helping to ensure that values acceptable to the public are utilized in the judgments. Scientists with extreme values, such as those for whom any regulation is too costly, or those for whom no risks whatsoever are acceptable, can be kept from influencing policy. Their advice can be discounted, not for their lack of competency or expertise, but for the lack of fit between their values and the polity's. When deciding upon a course of action affecting a democratic populace, it would be helpful to know what errors, and which consequences of error, an expert finds acceptable. Knowing what the expert is willing to risk in the face of scientific uncertainty will help produce better informed decisions by the officials acting on that advice, officials who can be held accountable (as they should be) for accepting the advice as a basis for action.

Making explicit the bases for judgments, including values, would involve a change in the traditional norms for science in policy and how they are put into practice. Happily, such a change could be assisted by the ongoing emphasis on making explicit the basis for judgment in risk assessment. For example, in the EPA's 2005 *Guidelines,* in addition to discussions of judgments and degrees of protectiveness, there is a consistent call for risk assessors to make judgments, and their bases, as clear as possible. For the weight of evidence narrative, risk assessors should be explicit about the full range of data available, assessments of the data's quality, "all key decisions and the basis for these major decisions," and "any data, analyses, or assumptions that are unusual or new to the EPA" (EPA 2005, 2-50). For risk assessments as a whole, the *Guidelines* call for "choices made about using data or default options in the assessment [to be] explicitly discussed in the course of the analysis" (ibid., 5-4). Such explicitness should also apply to assessments of sufficiency for departing from default inference guidelines, or for applying the guidelines themselves, including the case-specific social and ethical issues that shape such judgments.

Thus, the inclusion of a full range of social and ethical values need not threaten the integrity of science in policy or the production of risk assessments. While detached objectivity is required to avoid junk science, what of the other aspects of objectivity? Rather than ensure that a baseline requirement for sound science is met, the presence of objectivity's other aspects can increase our confidence in the reliability of science to be used in policy. For example, science that exhibits manipulable objectivity should increase our confidence that policy interventions based on that science will succeed in reaching their goals. Science reflective of interactive objectivity has been vetted by large and diverse (albeit relevant) groups of scientists in intensive discussion, and thus likely passes muster well beyond the baseline of requisite integrity. Finally, value-neutral objectivity would be an asset if the values at stake are wide ranging and contentious, and if the policymakers wish to take a middle-of-the-road approach to a problem. Then, scientists whose commitments fall in the middle of the value spectrum are most likely to produce judgments appropriate to the contentious context. In other contexts, where values are more bifurcated or some unacceptable values are present in the range, value-neutrality would be less of an asset. In these cases, a more deliberative approach, to be discussed in the next chapter, would be more appropriate.

Conclusion

Social and economic case-specific considerations have a necessary role to play in risk assessments. The need for some judgment in risk assessment cannot be escaped by developing inference guidelines. One cannot rely upon rigid inference guidelines without ignoring novel or developing science, and one cannot use all the available science without some case-specific judgments. The NRC in 1983 suggested that some considerations and not others be used to make the needed judgments, but all relevant social, cognitive, and ethical considerations should be used (in the indirect role) if the full responsibilities of scientists are to be met. Considering the importance of uncertainty in a specific case, or determining whether the evidence is sufficient to depart from an inference guideline, is a crucial part of responsible scientific judgment, and should include the consideration of the specific consequences of error.

Allowing case-specific social and political values in risk assessment need not pose a threat to the integrity of risk assessment. Integrity is maintained by constraining the values to an indirect role, by holding to detached objectivity. Considerations such as the seriousness of potential harms posed by a substance (or process), or the economic importance of a substance (or process), should not serve in a direct role in scientific reasoning for risk assessments. These considerations, no matter how potent, cannot be valid reasons for accepting or rejecting claims about the risks posed, nor can they provide evidential support for such claims. No matter how science's role in policy is conceptualized, this prohibition must be maintained, or the very value of science to the policymaking enterprise is undermined. These considerations can only help to assess the importance of uncertainty surrounding the claims made and to help assess whether the available evidence provides sufficient warrant.

In addition to keeping values to the indirect role in risk assessment, scientists should strive to make judgments, and the values on which they depend, explicit. Because policymaking is part of our democratic governance, scientific judgments made with these considerations have to be open to assessment by decisionmakers. Scientists should be clear about why they make the judgments they do, why they find evidence sufficiently convincing or not, and whether the reasons are based on perceived flaws in the evidence or concerns about the consequences of error. Only with such explicitness can the ultimate decisionmakers make their decisions based on scientific advice with full awareness and the full burden of responsibility for their office.

CHAPTER 8

VALUES AND PRACTICES

SCIENCE, EVEN SCIENCE TO BE USED IN PUBLIC POLICY, should not be value free. Scientists must make judgments about the acceptability of uncertainty, and these judgments require a range of values, including ethical and social values where relevant. The integrity of science depends not on keeping these values out of science, but on ensuring that values play only acceptable roles in reasoning. The direct role for values is acceptable only when deciding which research to pursue and, in a limited sense, how to pursue it. Once the work is under way, scientists must keep values to an indirect role, particularly when making judgments about which empirical claims should be made on the basis of available evidence. Values are not evidence, and should not take the place of evidence in our reasoning. But values are essential to meeting the basic moral responsibility to reflect on consequences of error.

Because of the need for democratic accountability in science for policy, the judgments made using these values should not be clandestine. Scientists need to make explicit the values they use to make these judgments, whether in their own work or when drawing on the work of others to complete risk assessments. One might be concerned that requiring scientists to be clear about all the considerations that go into their judgments, including the values that shape their judgments about the sufficiency of evidence, would make risk assessment documents far too complicated and unwieldy, while placing a heavy burden on scientists. In making the values explicit in all the judgments, the advice may become opaque. Surely there must be some easier way to proceed, which would allow scientists to not have to trace every judgment back to its source in the final documents, making every consideration explicit. At the very least, scientists need not do all this work alone.

There are two general ways to assist scientists in making the necessary judgments: (1) one can help the scientists make those judgments as the need arises throughout a study, and (2) one can decide prior to such judgments which particular values should be used to shape the judgments. For both of these approaches, greater public involvement would be beneficial. The public can help direct scientific studies (or the syntheses of studies such as risk assessments) that are to be used to make policy, assisting with the judgments as they occur throughout the process. The public can also engage directly with the social, economic, and moral values at issue in policy, deciding on a more general level what the value trade-offs should be. Experts can then use these value positions in their judgments during a particular risk assessment.

I am not the first to suggest that increased public involvement in the science and policy process would be beneficial. At least two reports from the 1990s suggested the same thing: the National Research Council's *Understanding Risk* (1996) and the Presidential/Congressional Commission on Risk Assessment and Risk Management's report (1997a, 1997b). Both emphasize the need for greater public involvement in assessing risk, particularly for the framing of risk assessment questions. While the need for public involvement in framing risk assessments is crucial, it does not capture all of the ways in which public involvement is important. In particular, if one adds a concern over judgments on the sufficiency of evidence, the need for public involvement in what the NRC calls "analytic-deliberative processes" becomes even more apparent. The NRC's analytic-deliberative process holds particular promise for incorporating the public into what has been the realm of scientists and technocrats. Conceiving of these processes in the NRC's manner can help scientists and risk assessors meet their responsibilities to consider the consequences of error, to include all relevant values in their proper roles, and to do so in a democratic and transparent manner. The NRC's conceptual framework also suggests an approach that allows the public to participate in risk assessments in a way that ensures its involvement poses no threat to basic scientific integrity.

Before looking at the NRC's report, a potential worry about increased direct public involvement with science in policy should be addressed. Allowing for direct public participation in risk assessment may seem to bypass the lines of democratic accountability discussed in the previous chapter. How can the public directly affect the risk assessments on which policy is based without undermining the legal lines of public accountability through which our democracy functions? It must be noted, however, that those lines, from citizens to Congress to agency officials, can be quite tenuous in practice. It would be rare for citizens to vote a representative out of office because she or he failed to properly chastise an agency official for an undesirable or problematic regulatory decision. Because of the tenuous nature of agency accountability, we have some reason to shift away from the strict representative democracy model this line of accountability embodies, to a more participatory model of democracy, where more direct citizen involvement in regulatory decisions is desirable.[1]

Daniel Fiorino (1990) notes three distinct reasons for such a shift: (1) the normative concern over democratic accountability—the representative model offers only weak democratic accountability, (2) the substantive concern over increasing the quality of information in regulatory processes—a participatory approach can assist policymakers in gaining a better understanding of unique local conditions relevant to the decisionmaking, and (3) the instrumental concern over increasing the acceptability of the final decision among the citizenry likely to be affected—the more involved the affected public is from the start, the more likely the final outcome will be uncontested (Fiorino 1990, also discussed in NRC 1996, 23–24). Yet the participatory model cannot take over the regulatory process completely, as the legal lines of authority, and the lines of accountability through representative democracy, must also remain in effect. Accountable officials must still make the final regulatory decision.

Happily, because science advice is just that—namely, advice—direct public involvement in risk assessment and other forms of science advice need pose no threat to the traditional lines of accountability. Risk assessments and other scientific summaries generated by partnerships between experts and interested stakeholders from the public have no special legal authority, and thus need not impinge on the traditional lines of accountability. Even if the advice may be more authoritative because of the public's involvement with its generation, the final decision still rests with an accountable public official.[2] Thus, it is in risk assessments that the public can have a truly participatory role without disrupting the traditional lines of accountability in representative democracy.

Public Participation in Risk Assessment: Analytic-Deliberative Processes

How can the public play a constructive and positive role in risk assessment? In 1996, the NRC reconsidered the framework for science in policymaking in the report *Understanding Risk*. Initially, the NRC committee was charged with examining problems with characterizing risk for decisionmakers (the final stage of risk assessment according to the 1983 NRC report discussed in the previous chapter, following hazard identification, exposure assessment, and dose-response analysis). In reflecting upon the issues with risk characterization, the committee found that an examination of the risk assessment process as a whole was needed to address the concerns raised with risk characterization. By rethinking the process of using science to regulate risk, the 1996 NRC committee developed a novel way to think about risk assessment. The committee described the process as an "analytic-deliberative" one, in which deliberation and analysis play mutually supporting complementary roles, and the process cycles repeatedly between analysis and deliberation.[3] By describing the process in this way, the NRC proposed a new framework for thinking about science and policy. The framework of "analytic-deliberative processes" recognizes a pervasive need for judgment and the resulting need to consider relevant values in making these judgments. In addition, it works to keep values from playing illegitimate roles in the process, and opens the door for more public involvement in risk assessment.

To appreciate the change in the perspective presented in the 1996 NRC report, it is useful to contrast it with the 1983 report. As defined in that earlier report, risk characterization was the stage where the risk assessor summarized the science in such a way as to present a useful estimate of the risk posed by the issue in question (NRC 1983, 20; NRC 1996, 14). In other words, the characterization of risk occurs "when the information in a risk assessment is translated into a form usable by a risk manager, individual decision maker, or the public" (NRC 1996, x). After completing the hazard identification, exposure assessment, and dose-response analysis, each structured by inference guidelines as needed, the risk characterization was to put all this information together for decisionmakers. The standard view of risk characterization was one of summarization or translation of science into a usable form, with "no additional scientific knowledge or concepts" and a "minimal amount of judgment" (ibid., 14). Once the risk characterization was complete, decisionmakers (perhaps in conjunction with the interested public) could begin the risk management process, where the difficult political decisions were concentrated.

Analyzing the problems with risk characterization in 1996, the NRC found that mere translation or summarization of science was inadequate. One of the most serious problems with risk characterization addressed by the NRC was the need to have a useful and appropriate risk characterization. As the report notes, "A carefully prepared summary of scientific information will not give the participants in a risk decision the understanding they need if that information is not relevant to the decision to be made. It is not sufficient to get the science right; an informed decision also requires getting the right science, that is, directing the scientific effort to the issues most pertinent to the decision" (ibid., 16).

Ensuring that the risk characterization "adequately reflects and represents existing scientific information" is not enough (ibid., 32). The issues relevant to the decision context must be addressed in the risk characterization. To achieve this, more thought is needed on how to get risk analysts to ask the right questions at the beginning of the risk regulation process. The NRC argued that instead of thinking of risk characterization as an activity occurring at the end of risk assessment, the risk characterization should instead "determine the scope and nature of risk analysis" (ibid., 2). In other words, risk characterization should be the process that formulates the initial questions to be addressed in the risk assessment, concerning itself with issues such as which hazards and which outcomes to consider. By addressing how the questions are initially formulated, the NRC argued for a broader understanding of risk characterization.

Because of these concerns, the NRC came to think of risk analysis as an "analytic-deliberative" process. This phrase takes on a very specific meaning in the report. "Analysis" is defined as the use of "rigorous, replicable methods, evaluated under the agreed protocols of an expert community—such as those of disciplines in the natural, social, or decision sciences, as well as mathematics, logic, and law—to arrive at answers to factual questions" (ibid., 3–4). It may help to think of analyses as processes that inherently meet the criteria for procedural objectivity, as discussed in chapter 6. In contrast, "deliberation," is defined as "any formal or informal process for communication and collective consideration of issues" (ibid., 4). Deliberations should strive for as much interactive objectivity as possible, with open assessment of reasons and inclusion of diverse relevant perspectives. In the risk regulation process, both analysis and deliberation are needed: "deliberation frames analysis, and analysis informs deliberation" (ibid., 30). The relationship between analysis and deliberation should be ongoing and mutually influential. One does not complete an analysis and only then turn to deliberation (or vice versa). Instead, the process is "recursive," and the two aspects of the process "build on each other," as "analysis brings new information into the process," and "deliberation brings new insights, questions, and problem formulations" (ibid., 20).

To emphasize the pervasive need for deliberation, the NRC report spends considerable effort in describing points of judgment important for risk analysis. They include how the problem is conceived (for example, is fairness in the distribution of risk considered), which outcomes are examined (such as, what are the ecological ramifications, or the effects on future generations?), and which measures of risk are to be used (for example, deaths or years of life lost) (ibid, 38–50). All of these are judgments that should be considered in the initial framing of the risk analysis, and in which presumably the public has some interest. Some preliminary analyses may help inform these judgments, reinforcing the need for recurring interaction between analysis and deliberation. But deliberation over which choices to pursue is essential to making these judgments with due care.

In addition to judgments made early in the process, the NRC also recognized the need for judgment later in the process, particularly when deciding how to characterize uncertainty in the final characterization of risk (ibid., 66–67, 106–16). Overly technical uncertainty analyses can be opaque; ignoring uncertainties can be misleading. A deliberative judgment on how to proceed must be made. As the NRC noted, "Because of the power of formal analytical techniques to shape understanding, decisions about using them to simplify risk characterizations or to capture uncertainty should not be left to analysts alone, but should be made as part of an appropriately broad-based analytic-deliberative process" (ibid., 117). In doing analyses, as with any procedurally objective process, once the analysis is under way, judgment should be minimized by virtue of the nature of the process. But the initial selection of which analytic process to pursue is crucial. Analytical processes are themselves not value free, even if procedurally objective. Deliberation is needed to ensure that the appropriate values shape the decision of which analyses are performed and utilized. The potential need for deliberation over issues raised in a risk assessment engenders the possibility for the public's ongoing input into risk assessments and suggests new ways in which the public could legitimately shape them.

However, it would be overly simplistic to think that analysis remains the purview of the expert and deliberation the purview of the public. Deliberation is recognized to include discussion "across different groups of experts" and "between experts and others," in addition to deliberations among public groups interested in the issue (ibid., 30). Deliberation is also recognized to be an important aspect of the internal functioning of any scientific community, as exemplified by the use of peer review in science (ibid., 74). It is needed wherever judgment is needed. Analyses, on the other hand, as defined by the NRC, are processes that do not require judgment. Even though analyses may require "contributions from many, diverse disciplines," the need for judgment should be minimal (ibid., 24). The NRC emphasizes this when they write that "good scientific analysis is neutral in the sense that it does not seek to support or refute the claims of any party in a dispute, and it is objective in the sense that any scientist who knows the rules of observation of a particular field of study can in principle obtain the same result" (ibid., 25). Thus, analysis should be detached (capturing the idea of "neutral" in the NRC report) and procedurally objective, in that it should not matter who completes the analysis. As the need for judgment in analysis is kept to a minimum, the need for deliberation at points of judgment elsewhere in the process, particularly in the selection of analytic techniques, increases.

There is one place of judgment, however, that is overlooked by the 1996 NRC report. The assessment of the sufficiency of evidence, or the acceptability of uncertainty, is not discussed in the document. When this point of judgment is included in one's consideration of risk analysis, the endemic need for judgment in the process is even more pronounced. When deciding upon the sufficiency of evidence to support an empirical claim, or deciding whether an analytic technique or a simplifying assumption is sufficiently accurate, a consideration of the consequences of error is needed that must include social and ethical values. These are points of judgment that would be greatly assisted by deliberative processes, particularly those that brought in the appropriate range of the public's interests and values. The inclusion of this point of judgment underscores the need for an analytic-deliberative approach to risk analysis.

The analytic-deliberative framework for the understanding of values in science and policy is illuminating. Values, even case-specific ones, are not kept to the later political portion of the decision process (that is, risk management). Instead, values are needed in the initial deliberations that are performed to frame the subsequent analyses. As described in chapter 5, there are two general roles for values in science. When one is deciding upon which questions or methodologies to pursue, a direct role for values is often needed. The initial framing of a risk assessment allows for this role, as values direct the attention of risk assessors. An indirect role for values may also be needed in the selection of analytic techniques. Once analyses are under way, the acceptable role for values shifts to the indirect role only. Analyses by definition require minimal personal judgment, following the standards of the experts' fields. When judgment is needed in analyses, it arises where there is uncertainty that must be weighed. Here the consequences of error become important, and must be discussed in a deliberative manner.

Such judgments, both at the beginning of and during analyses, should be guided by values relevant to the case, values discussed in a deliberation. The deliberation, if not already begun before the need for judgment becomes apparent, should involve more than the small expert group (as an expert group may have skewed values), bringing into the process members of the public. This bolsters democratic accountability of expert judgments, but does not threaten the integrity of expert work, as the deliberations cannot suppress analyses or require that they produce particular answers (an inappropriate direct role for values). Values should not determine the outcome of the analyses, but only shape judgments where needed. For example, although the public may be concerned over particular health risks to children, thus focusing analytic attention on this aspect of risk, subsequent analysis may show such risk to be less than anticipated, and other risks may appear more significant. Or, once an analysis is under way, experts might feel the need for judgment when faced with uncertainty on how to proceed. The experts should consult with a broader group in a deliberative process to determine which choice to make among the analytically acceptable but disparate options in front of them. The group's decisions on that judgment thus shape the analysis, but do not corrupt it. Completed analyses then serve as the basis for further deliberation on courses of action.

There is no "value-free" part of the analytic-deliberative process; case-specific values play a role throughout, and that role can be a legitimate one. If experts maintain authority over the analyses, and judgments shaping those analyses are kept to the indirect role, values cannot block the work of the experts. When experts must make choices among analytically acceptable options, the experts should turn to the broader group to assist with the judgments needed to inform the choices, allowing ethical and social values to weigh the possible errors of different approaches. The public nature of expert judgments, accountable to all interested parties, keeps them from becoming sources of contention or accusation later on. This can greatly bolster public trust in risk assessments. In addition, making such judgments clear is not a threat to scientific integrity as long as the role of values is kept to their legitimate forms, not suppressing or inventing evidence. Then the appropriate norms for values in scientific reasoning are upheld.

Analytic-Deliberative Processes in Practice

This might seem fine in theory, but can it work in practice? Can experts and the public fruitfully work together to produce risk assessments that engender trust on all sides? Happily, there are some examples of analytic-deliberative processes in the literature on public involvement with policymaking. (The processes that most closely model the analytic-deliberative framework are often called "participatory research" or "collaborative analysis" or some combination thereof, rather than "analytic-deliberative processes.") In examining such examples, we can see both the promise and the practical barriers to implementing the analytic-deliberative framework.

One of the most striking examples comes from Valdez, Alaska. In his detailed study of disputes concerning the marine oil trade in Valdez, George Busenberg (1999) compares a period in the early 1990s characterized by "adversarial analysis" (that is, competing experts used in lengthy and unresolvable disputes marked by a lack of trust) with a later period characterized by "collaborative analysis." Busenberg's account of collaborative analysis in Valdez is both intriguing and promising, reflecting what can happen in a successful analytic-deliberative process, and how it can get past the problem of dueling experts. The two opposing groups, the oil industry and a community group called the Regional Citizen's Advisory Council or RCAC (formed in 1989 after the Exxon Valdez oil spill), had a history of distrustful and confrontational relations, including scientific experts marshaled on either side. By 1995 they both seemed to realize the impasse to which this generally led, and resolved to find a way around these difficulties. The dispute at that time centered around what kind of tug vessels should be deployed in the Prince William Sound to help prevent oil spills. Instead of doing competing risk assessments to influence the policy decision, RCAC, the oil industry, and the relevant government agencies decided to jointly sponsor and guide the needed risk assessment (Busenberg 1999, 6).

The risk assessment proceeded with a research team drawn from both industry and RCAC experts. The project was funded by the oil industry, RCAC, and the government agencies. The steering committee had representatives from all of these groups and met regularly (fifteen times) to direct the study. Not surprisingly, members of the steering committee learned much about the intricacies of maritime risk assessment, which led to increased appreciation of the expertise involved. More interestingly, the research team of experts found the guidance of the steering committee to be very helpful. As one expert stated, the process "increased our understanding of the problem domain, and enabled us to get lots of data we didn't think was available, [and] . . . the assumptions were brought out in painful detail and explained" (quoted in ibid., 7–8). The additional data were needed when the committee decided that "existing maritime records were an insufficient source of data for the risk assessment models" (8). Clearly, the risk of error was thought to be too great, and the value divisions too large to bridge.

One way to deal with such a situation is to reduce the risk of error. In this case, the steering committee assisted the researchers in gaining more detailed data needed to complete an adequate risk assessment, including data on local currents and weather patterns. Thus, not only did the stakeholders (the steering committee in this case) help the risk assessors determine which assumptions in the face of uncertainty are acceptable (given the potential consequences of error), they also helped determine that a certain level of uncertainty was altogether unacceptable, and that it required a further investment of time and resources to gather more data, with the aim of reducing the uncertainty. The final risk assessment was accepted as authoritative by all parties, and one of the new tug vessels was deployed in the sound in 1997 as a result (ibid., 8).

This example shows the promise of analytic-deliberative techniques when deployed at a local level to address contentious environmental issues. Where there are clearly defined stakeholders representing various public interests, they can collaboratively design, fund, and direct research that can resolve technically based disputes. Crucial to the process is an equal sharing of power among the parties, made possible in part by joint funding of the research. Stakeholders play a key role in directing the research and in helping to shape the scope and guiding assumptions of the analysis. The stakeholders' values are thus able to play their proper role at the appropriate points of judgment. In the Valdez case, the steering committee served as the key deliberative body, and the research team as the analytic body. Continual feedback between the two groups produced a trustworthy result, one that was not derided as unsound science, and that could be used as a basis for decisions without contention over the final risk assessment.[4]

The key question for the promise of analytic-deliberative processes is the extent to which they are feasible across a wide range of contexts. The Valdez case had near ideal conditions, with clearly defined stakeholders committed to finding a reasonable solution, adequate resources to pursue the approach, and policymakers willing to listen to the final outcome. Lacking any one of these components can doom the process. Thus, it is doubtful this sort of intensive analytic-deliberative process can be a universally applicable approach to science in policy issues. Although there are clearly limitations to a full-fledged analytic-deliberative process, precisely how to best pursue the potential benefits of such processes and where the bounds of the benefits lie is not yet clear. What is needed now is more experimentation with analytic-deliberative processes, to assess their potential to assist scientists with these difficult judgments and to assess their ability to move us past the chronic problem of policy hamstrung by feuding experts.[5] Just as we once experimented with ways of achieving effective science advising in the federal government and built upon small successes, experimentation and then broader implementation is needed to find our way here.

Some lessons have already been learned. It is clear that one needs to have policymakers fully committed to taking seriously the public input and advice they receive and to be guided by the results of such deliberation. When that commitment is missing, participatory processes can erode the relationship between policymakers and the public. For example, in an attempt at an analytic-deliberative process at Hanford, Washington, concerning groundwater contamination at the former plutonium production site, the relevant authorities (the Department of Energy) began a collaborative process with local stakeholders for directing relevant risk assessments, but then largely ignored or disavowed the process (Kinney and Leschine 2002). This led to frustration among the stakeholders and a worsening of an already difficult relationship with the DOE. Trust was undermined rather than bolstered, and the stakeholders were embittered, having committed to a process that ultimately was fruitless.

Another necessary component for successful analytic-deliberative processes is a public that is engaged and manageable in size, so that stakeholders can be identified and involved.[6] This aspect is particularly difficult if the issue to be addressed is national or global in scope, rather than local. Some issues of a more global nature, such as climate change, face the challenge of identifying an acceptable set of stakeholders to guide a risk assessment process. Given that hundreds of scientists across a range of disciplines are needed to produce the Intergovernmental Panel on Climate Change reports, it seems unlikely that a manageable number of stakeholders could be identified for such a project. In these sorts of cases, a participatory research approach appears less feasible, and alternative approaches are called for. One possible approach is to ask local groups across a range of relevant contexts to each perform their own participatory risk assessment and to feed their results into decisions at the state or national level. Something similar has been attempted with the national issue of how to best dispose of outdated chemical weapons—a national issue with decidedly local implications, as dismantling or incineration facilities would be sited locally (Futrell 2003). Or it may be best to focus the public on the values at stake, and the appropriate trade-offs, using techniques I discuss in the next section.

Finally, the process needs to be managed carefully, so that trust and integrity are maintained on all sides. If analytic-deliberative processes become beholden to particular interests at the outset, they will ultimately revert to the feuding expert model, with opposing interests touting their analytic-deliberative process as more reliable than their rivals'. This would defeat one of the most attractive pragmatic benefits of these processes, of providing a way to circumvent stalemated debates rather than serving as yet another point of dispute.

In sum, we should not expect the analytic-deliberative process to be a workable model for all situations, particularly in the ideal form of direct and intensive public involvement with risk assessment or science advice. Yet even in cases where an ideal analytic-deliberative process does not appear feasible, the public can still play a key role. Rather than attempt the difficult task of direct and ongoing involvement, it may be possible to enlist the public in narrowing the range of acceptable values used for making judgments in a risk assessment. If experts then use the narrowed range explicitly, with those values properly guiding their judgments, some democratic accountability could be maintained. Simply put, the public could assist scientists by clarifying the appropriate values to use in their judgments.

Working on Values

When policymaking is likely to affect a large and widespread population, it may not be feasible to have scientific experts working in close contact with public stakeholders because the public is too large and geographically widespread. In these cases, the public can work on the values directly, with some minimal scientific input, rather than pursue the intensive interactions described above. If consensus on the appropriate weighing of key values central to particular issues can be achieved, experts making judgments about the acceptability of uncertainty can (and should) use that consensus valuation. This keeps the scientists from imposing their values on an unwitting public, and allows the public to have their say on how the judgments should be shaped. Achieving such consensus requires social practices that improve our current anemic discourse about values.

The period from 1995 to 2005 has witnessed an increased interest in practical ways to generate the necessary dialogue among members of the public, as social scientists have begun developing methods that would allow for constructive deliberation about values centered on particular issues.[7] One particularly interesting approach, developed in Denmark and exported to other nations, is the consensus conference. Consensus conferences recruit average citizens for an intensive deliberative experience, including education on an issue, group deliberations, further questioning of expert panels, and ultimately a drafting of a consensus statement on the issue (Joss and Durant 1995).

In general, between one to two dozen citizens are recruited, usually representing a range of social and economic backgrounds. Some randomized searching is usually incorporated into the recruiting process to ensure an appropriate distribution. The citizens are not "stakeholders" in the usual sense of people with a special interest in an issue. Rather, they are members of a vaguely interested public, people whose lives will likely be affected by an issue, but who have not been involved with the issue previously. These citizens serve as surrogates for the general populace, and, during the conference, learn the intricacies of the particular policy issue at stake. They usually attend several talks by a range of experts, which serve as a starting point for their discussions. They then meet without the experts to discuss the issues and to formulate more penetrating questions for the expert panel. Upon hearing the answers to those questions, the citizens deliberate for a day or more to formulate a consensus statement. The whole process can take less than a week (although it can be extended over several weekends). The strengths of the approach are well documented: the citizens involved learn a great deal about an issue; the group deliberation is often reflective and revealing of deep issues and addresses value-based issues with nuance; the consensus document often reflects well what the citizenry as a whole would think after similar depth of exposure (Sclove 2000). Because of the well-informed nature of the deliberations, consensus statements on the relevant value trade-offs can serve as effective guides for scientists when making judgments involving similar trade-offs.

Consensus conferences can serve as useful examinations of values for particular issues, especially when the issue is of national or global scale, making a participatory research approach infeasible. They are not without their difficulties, however. Because of the small size of the public group selected, one can question whether they are adequately representative of the public. However, we would do well to keep in mind that the guidance that consensus conferences provide is several steps removed from final decisionmaking authority. Recommendations arising out of consensus conferences are at most advice to policymakers, on the same level as advice from scientists, which currently is held to no democratic standards. More frequently, consensus conferences provide advice to advisors, and thus inject some democratic sensibility into a process that would otherwise have none. Unless the recipients of the advice have specific reasons to be concerned about the particular process, they should give the results some weight. And as with participatory research approaches, we can look forward to much more experimentation and refinement of process. A consensus conference dealing with telecommunications issues in the United States, for example, had instructively mixed results (Guston 1999). Although the conference produced no impact on policy (it was completely ignored by policymakers), it did impact positively the way people viewed the prospects for such modes of public participation in general.

While we experiment with social mechanisms to achieve a robust dialogue and potential consensus about values, philosophy may also be able to improve our discussions of values. Admittedly a long shot, some philosophical techniques could be helpful. I emphasize techniques over theory because there is not one ethical theory that can be applied to all decision contexts, producing the "correct" values that can then be used to guide judgments. There are deep divisions among philosophers about the correct ethical theory (or whether there even is one), and thus it would only draw members of the general public unwittingly into long-standing philosophical disputes. Instead of an application of a particular theory to a particular problem, the contribution philosophical ethics might make is in helping to structure thought experiments that could further deliberations about value questions.[8]

There are at least three general approaches to thinking about values that could be of use here. Each allows reflection on one's values through a thought experiment, and are thus a kind of test of the value positions. These approaches can help citizens think through the implications of their value positions, and help them test the extent and nature of their commitment to those values. The thought experiments they generate may even produce shifts in values, potentially pushing extreme positions toward a consensus, although this is likely to be a culturally slow process. Three general methods for generating ethical thought experiments are: (1) universal reflexivity (or universalizing), (2) consequentialist calculus, and (3) analogous reasoning.

For the first, universal reflexivity, the experimenter asks questions such as, if others held the same values and acted accordingly, would the world be an acceptable place?[9] If the answer is no, the acceptability of the values is dubious. Or, the experimenter could ask whether others would hold the same values regardless of irrelevant details about their own existence.[10] (Of course, what counts as a relevant detail becomes a crucial issue.) The test of universal reflexivity helps us examine if the value we hold is simply special pleading for our particular interests, or whether it has a broader basis.

The consequentialist calculus, on the other hand, considers each individual's desires as important, and attempts to maximally satisfy the desires of the most number of people possible. Thus, it asks, if you act on the basis of a certain value, will you produce the greatest happiness for the greatest number? Arising from utilitarian theories of ethics, the consequentialist calculus is similar to cost-benefit analysis. In addition to testing the "goodness" of a choice in this way, the approach allows for a deeper examination of certain issues. In constructing the details of a consequentialist calculus, a number of important valuation issues rise to the surface. For example, the following questions arise: Is providing good to a large number of people better than providing more good to a smaller number of people? How are goods among individuals to be compared? Do goods in the future count as much as goods in the present? Addressing these questions can clarify the values at stake for a decision.

Perhaps most useful for improving deliberations about values is good analogical reasoning. Analogical reasoning constructs a case parallel to the issue at hand, thinks through that case, and then applies the reasoning in that case to the original issue. The power of this technique lies in the construction of a case with the same logical structure as the original, but without all of the emotional overtones of that case. If the parallel is a good one, it can reveal inconsistencies in one's value structure that most people find intolerable. As a result, it can generate shifts in value systems. One particularly powerful example of this kind of reasoning comes from Judith Jarvis Thomson (1971). In her example, a world famous violinist lies grievously ill and in need of another person for life-support. Music fanatics kidnap a random person off the street and force a doctor at gunpoint to hook the kidnapped person up to the violinist to act as the life-support system. Once connected, any severance would likely kill the violinist (who has authorized none of this nefarious activity). Suppose you are the kidnapped person. You wake up attached with tubes to the violinist, informed that without you, the violinist will die. (The fanatical gunmen have fled the scene.) Are you obligated to remain there for any number of months until the violinist improves?

Thomson uses this example to generate careful reflection about abortion in cases of rape or incest. In both cases, the being on life-support is an innocent, ending the life-support will kill the innocent, and others used violence to force someone to serve as a life-support system. While it may be kind to remain in the hospital, the question is whether the person serving as life-support should be considered morally or legally obligated to do so. The example strips away the cultural trappings of motherhood and the blaming of rape victims to focus on the key moral questions, and thus it serves as a fruitful analogy, which can then be tweaked to explore the extent of one's moral commitments.

Each of these general reasoning processes can enrich deliberations about values. If such deliberations can produce a consensus about value trade-offs, or even narrow the range of acceptable values, the judgments scientists face can be made both less contentious and more accountable. We have achieved consensus about values in the past, deciding as a society, for example, that racism and sexism are wrong. No public policy can be explicitly based on such values (even if some policies help perpetuate inequities in these areas), and it is considered important to ferret out implicit reliance on sexism or racism. These achievements were not easy, taking a great deal of moral and cultural struggle, requiring the efforts of many people over decades. Many of the value questions central to science in policy need similar attention and effort, such as how we should consider impacts on future generations and to what level of environmental quality do our citizens have a right. Getting a handle on these philosophical questions is of utmost importance to the furtherance of public policy.

None of these approaches, from participatory research to consensus conferences to long-term philosophical efforts, alleviates the need for scientists to be reflective about the judgments they make in their work and the values that inform those judgments. It is unlikely that any of these mechanisms could fully remove the burden from scientists to consider the consequences of error, even in a particular, limited context. Thus, even in a full-fledged analytic-deliberative process, scientists need to be aware of the judgments required to structure the relevant analyses, so that those judgments can be shared with the participating stakeholders. In addition, the scientists' expertise would likely be needed in any deliberative processes, particularly for showing the range of implications at stake and thus the scope of potentially relevant value considerations. In other cases, when a consensus conference provides some guidance on crucial value tradeoffs, it would still be up to the scientists to put that guidance to use when making judgments. Making the presence and nature of these judgments as explicit as possible will be essential for acceptable democratic accountability. As science takes on more importance in decision contexts, these aspects of scientific work will take on an increasing prominence.

Scientists, however, should welcome the opportunities that such challenges present. Scientists in the second half of the twentieth century repeatedly lamented the lack of scientific literacy among the public. Both collaborative analysis and consensus conferences provide scientists an opportunity to help a segment of the public (admittedly limited) become more aware of what their work entails, and indeed of the very nature of scientific endeavors. In these contexts, scientists have an engaged, perhaps captive, if potentially critical, audience. As scientists come to understand the full implications of their increasingly public role, they need to be open to genuine public input on how to best fulfill that role. One cannot claim to be publicly accountable but unmoved by public criticism. Nor can one claim autonomy from public accountability if one wields power in the public realm. In short, just as scientists want the public to listen to them, they will, in turn, have to learn to listen to the public.

Values, Diversity, and Science

Is direct public involvement in the processes that produce science advice really necessary? There is, after all, some diversity in the scientific community, as well as calls for increasing that diversity. For example, both Helen Longino and Miriam Solomon emphasize the importance of a socially diverse scientific community for an optimal critical discourse in science (although with different emphases) (Longino 1990, 2002; Solomon 2001). Without having people of diverse backgrounds involved in science, the possibility increases that unexamined and potentially pernicious hidden assumptions will have undue influence in the framing of scientific hypotheses. Perhaps the diversity needed for better epistemic practices is sufficient for vetting the values in scientific judgments embedded in scientific advice.

There are several reasons, however, for thinking that this approach to the issue is likely to be insufficient. First, although the diversity of the scientific community has been happily increasing since the 1980s, it still falls far short of the diversity found in national demographics, much less international demographics. Second, many of the most central values scientists hold are those they developed while training to be scientists, and this alone can create divergences between the scientific community and the general public. We all tend to think that our chosen work or profession is important for society, or to assume it will be beneficial to the world. With this "insider" perspective, it can be more difficult to take seriously the risks embedded in one's work and to acknowledge potential pitfalls. Thus, even a demographically diverse scientific community may still hold values that clash with the broader public's values.[11]

Of particular relevance to science in the policy process are the values that shape what counts as sufficient evidence for a scientific claim. Although disciplinary standards may be set at one level, reflective of a concern to not admit false statements into scientific theory, the broader public could well be equally and legitimately concerned with not overlooking potentially important, if as yet undefinitive, evidence. In such cases, the values used to make judgments of sufficiency need to be clear and explicit, so that scientists, policymakers, and the public understand the basis for a claim. Finally, even if a scientific community has values that mirror those of the general public, the involvement of the public assists with making those values explicit, and for creating contexts where the judgments on which they rely can be examined carefully. Making the judgments explicit is important both for scientific epistemic practice and for democratic transparency and accountability.

While there are thus good reasons to be skeptical that a scientific community, even if fully diverse, can do without public involvement, scientists are still likely to be wary of such involvement, for it will not completely remove the burden of reflecting on judgments and will add procedural complication and complexity. So what are the potential positive benefits of public collaboration in science advice? First, as noted in the Valdez case above, having a public body steering the process can help scientists by requiring them to go over the assumptions and practices that guide the synthesis of evidence. As the researchers noted, this care and level of detail brought out points of judgment in explicit detail. This allows researchers to more carefully examine what they are doing and why. Second, the public body provides both a sounding board for the values needed to make the judgments and additional guidance on the judgments, thus assuring researchers that the values used are not inappropriate. While labor intensive, such practices can assure scientists that they are fully meeting their responsibilities. The shared effort can also improve the science from an epistemic perspective by increasing reflection on the practices and by gathering new data. Finally, for scientists unused to this level of reflection on their practices, having such a process can help them adjust to what it takes to meet their responsibilities. In short, the public helps the scientist through the necessary process.

Getting the public more involved with risk assessments and the issues surrounding particular policy problems presents new practical and theoretical challenges. We must figure out how to better engage the public, how to better address questions of values, and how to get experts and the public in a constructive dialogue. Happily, there is good research on which to build, as social scientists have already begun experimenting with and testing multiple ways to bring about fruitful collaboration. Hopefully, the arguments here will provide additional normative impetus to pursuing those avenues, even as they challenge the prevailing tradition of keeping the scientific community aloof from genuine public dialogue.

EPILOGUE

Reliance on the value-free ideal has produced something of a mess. Scientists have thought that any consideration of ethical or social values, particularly in the assessment of evidence, would undermine scientific integrity and authority. Yet one cannot adequately assess the sufficiency of evidence without such values, especially in cases where science has such a profound impact on society. Thus, a crucial source of disagreement among scientists has remained hidden, unexamined and unacknowledged. This has heightened the sound science–junk science disputes, as a lack of expert consensus often leads to charges of junk science on one side or the other. When junk science reduces to name calling, neither the integrity nor the authority of science is well served, nor is the policymaking that depends upon good science.

It is time to reject the value-free ideal and replace it with something better. The value-free ideal, in its most precise form, dates to the late 1950s. Ironically, it was forged by philosophers of science just when scientists were taking on an increasingly public role that belied the isolationist premise of the value-free ideal. Once this isolationist premise is discarded, the justification for the value-free ideal crumbles. With a full understanding of the public role of science, a new ideal becomes both possible and necessary.

The new ideal starts with a careful examination of the moral responsibilities of scientists. Because of the important public authority of science, the basic responsibilities to take care when making empirical claims, to consider the consequences of error of both making overly strong or overly weak claims, cannot be ignored or set aside by scientists. Although scientists can recruit assistance in shouldering this responsibility, they can never be completely free from it. We will always need scientists to interpret their data, to make clear statements about uncertainties, and to clarify what is at stake in our subsequent decisions. Scientists must make judgments to fulfill these roles, and with judgment comes responsibility.

Yet, the values that are a needed component of scientific judgment must also be constrained in how they function in science. Even in rejecting the value-free ideal, we would be foolish to allow values to serve in the same role as evidence throughout the scientific process. That would indeed undermine the value of science itself, its basic integrity and authority. However, we can retain the baby and throw out the bathwater by differentiating between the roles values play in science, and restricting values to the indirect role at key epistemic moments in science—particularly when assessing the strength of evidence for empirical claims.

Thus, scientific integrity consists in keeping values to their proper roles, not in limiting the values that have a role to play. Scientific objectivity rests in part on this distinction, as at least one component of scientific objectivity depends upon keeping distinct the roles values play in reasoning. There are other aspects to objectivity as well, each with its own ability to bolster our trust in a claim, and with its own potential pitfalls. Scientific claims can be checked against several aspects of objectivity to see which might be the most trustworthy; the complexity of objectivity is of great practical use. Yet the core of scientific integrity remains the restriction on a direct role for values to only appropriate stages, where values legitimately serve as reasons in themselves for a choice, such as in the selection of research projects.

With this key protection of scientific integrity, the understanding of how science should be used to make policy shifts. No longer should social or ethical values, even case specific ones, be held apart from an assessment of evidence. Such values are essential to the assessment. Instead, the values should only weigh the significance of the remaining uncertainty. And, equally important in a democracy, the values should be made as explicit as possible in this indirect role, whether in policy documents or in the research papers of scientists.

Making the values explicit will take some effort by scientists and policymakers. It may also be beneficial to bring the public directly into the process by allowing them to help make the needed value judgments. Research on social mechanisms to effect such participation has begun, and should be pursued with vigor. These efforts at public involvement will complicate the policy process, moving away from the linear model of facts first, value judgments later. But this is a necessary result of an honest assessment of the nature of scientific knowledge. Although science is our most reliable source of empirical claims, it is never certain, and must be open to being overturned by new evidence. The world can always surprise us, and when we cease being open to such surprises, we enter the world of genuine junk science.

On the shifting terrain of sound science, policy decisions still must be made. If values are openly acknowledged in their proper role, we will be able to see in retrospect how we might have made a decision that was later undermined by the gathering of additional evidence, but still understand how it was the best decision at the time, given our value commitments and available evidence. In an uncertain and complex world, this is the most we can ask.

NOTES

Chapter 1: Introduction

1. Gross and Levitt lump all critics of science together as "left." Many find this claim preposterous, given the views of some of the authors criticized by Gross and Levitt. See, for example, Berger 1994.

2. There is some disagreement over whether the editors of *Social Text* thought the essay was genuinely good or whether they published it simply because it was an attempt by a physicist to join the discussion. *Social Text* is "a non-refereed journal of political opinion and cultural analysis produced by an editorial collective," and thus should not be considered a mainstream academic journal, for whom peer review is central. (Editorial policy quoted in Boghossian 1998, 26; see also MacCabe 1998.) Christopher Hitchens (1998) reminds us of "two little-mentioned aspects of the case: first, the fact that at least one member of the *Social Text* editorial board does not believe the Sokal essay was a put-on; second, the fact that a conservative critic, writing in the *Wall Street Journal*, used the same essay to demonstrate the sort of rubbish that ST [*Social Text*] was inflicting on the 'public'" (44).

3. Consider, for example, the N-ray affair or the Piltdown Man. See Broad and Wade 1982, 107–25, for more details.

4. There is a venerable tradition of forgeries, but these seemed to be more aimed at out-and-out deception of everyone, in an attempt to make money or to gain prestige rather than to catch a colleague at sloppy scholarship. See Grafton 1990 for an engaging account of forged texts in the Western tradition. Grafton's assessment of the forger in the end is harsh: "Above all, he is irresponsible; however good his ends and elegant his techniques, he lies" (126).

5. Some have seen the Science Wars as providing ammunition for creationists and/or intelligent design theorists. With social epistemologist Steve Fuller recently acting as an expert witness on the side of intelligent design theorists, there seems to be something to this. But outside of this typically American and anomalous debate about science education, I see little impact of the Science Wars on the role of science in the United States.

6. The OMB has since crafted guidelines that would assist Federal agencies in meeting these standards. The OMB guidelines (OMB 2002, 2005) make clear how

difficult it is to meet the goals of the legislation. The OMB relies upon a narrow construal of integrity (defined as the lack of unauthorized tampering) and defines quality as just utility, integrity, and objectivity (OMB 2002, 8459–60). The key criterion remaining, objectivity, is assessed largely through peer review of documents, supplemented by reproducibility of analyses by others where relevant. Depending on the charge given to peer reviewers, peer reviews can be more or less helpful with this task. Unfortunately, the OMB instructs agencies to prevent peer reviews from assessing the acceptability of uncertainty in government documents (OMB 2005, 2669). Debate over how much evidence is enough, or how much uncertainty is acceptable, is a key source of contention in the sound science–junk science debates.

7. Concerns over the suppression of evidence by the Bush administration are raised in Union of Concerned Scientists 2004 and the August 2003 Waxman Report (U.S. Congress 2003). Similar concerns are raised from a different political angle in Gough 2003.

8. The value-free ideal for science, which will be historically traced and explicated further in chapter 3, holds that scientists, when reasoning about evidence, should consider only epistemic or cognitive values. For recent defenses of the ideal, see Lacey 1999 and Mitchell 2004. The arguments presented here in the ensuing chapters are attempts to adequately address the concerns they raise in their work.

9. Philip Kitcher (2001) addresses the problem of how society should decide which research agendas should be pursued. This is a difficult area, as one must assess the (potential) value of various possible pieces of knowledge one could acquire. This kind of assessment opens the door to comparing the value of that knowledge against other social goods, such as justice or economic prosperity. This line of inquiry also runs into problems of the freedom of inquiry, and how that freedom is to be articulated and what its boundaries are. These issues are outside the scope of this book.

10. Longino (2002) argues that knowledge itself is necessarily social.

11. See, for example, Wolpert 1992, 17–18, and Hoffmann 2003, although Hoffmann expresses concerns over relying on so-called "epistemic" values.

12. See also Margolis 1996, Slovic 1987, and Viscusi 1992, chap. 2, for more discussion on these aspects of risk.

13. It is doubtful that intelligent design, the most recent effort, can be considered science at all. I have not yet seen any competent predictions about the natural world that follow from the theory, and such prediction is a necessary part of science. As I will discuss further in chapter 5, the explanatory power of a theory is helpful, but predictive competency is essential.

Chapter 2: The Rise of the Science Advisor

1. Regrettably, a broader study of the rise of science advising across countries and cultures must await future work.

2. The Department of Agriculture encompassed the Weather Bureau, taken from the army in 1890, and the Forest Service, as well as agricultural research supported through land-grant universities.

3. As Dupree (1957, 334) notes, the NCB could not decide where to site its new lab, so the Naval Research Laboratory was not formally established until 1923.

4. Kevles 1995, 126, emphasizes sonar; Dupree 1957, 319, emphasizes tactics.

5. The budget for NACA increased to $85,000 in 1916 in response to the engagement of the United States in World War I (Kevles 1995, 108); in 1923 its budget was $200,000 (Dupree 1957, 334).

6. The NRB was renamed the National Resources Committee in 1935. To avoid confusion with the National Research Council, I will use the NRB acronym.

7. See Pursell 1979 for a detailed account of the creation and tenure of the NDRC and its successor, the OSRD.

8. For an amusing account of why the British report is known as the MAUD report, see Rhodes 1986, 340–41.

9. Another take of World War II as a watershed is captured by Mullins 1981: "Except for emergencies, the Federal government made no systematic use of scientific and technical advice on any significant scale until World War II" (5).

10. The Office of Naval Research (ONR) had a surprising, serendipitous origin; see Sapolsky 1979, 381–84, for the story. The ONR became the primary source of funding for academic scientists doing basic research from 1946 to 1950 (ibid., 384). Its importance to scientists of the period is underscored by Sapolsky's comment that "The golden age of academic science in America actually lasted only four years," the four years when ONR dominated research funding with little oversight (386). After 1950, ONR was under increasing pressure to fund projects that would serve the needs of the navy. Prior to 1950, it had adequate funds to distribute, but was not overseen by the navy to ensure the practicality of its funding decisions. As Sapolsky notes, "in 1950, the Navy rediscovered the Office of Naval Research," and scrutiny ended the "golden age."

11. See Kimball Smith 1965 for a full account of scientists and the struggle over the fate of nuclear power from 1945 to 1947. See Hewlett and Anderson 1962, chap. 14, for a detailed history of the 1946 Atomic Energy Act. See Lapp 1965, 87–100, for another account.

12. For a surprising recollection of *Sputnik*, see Bronk 1974, 120–21. Bronk recalls panic among the public, but also a desire among scientists to congratulate the Soviets for their achievement, as well as excitement for the research potential satellites presented.

13. See Lapp 1965, 113–30, on the role of the citizen-scientist in the 1950s concerning the debate over fallout for an example of this ethos in action.

14. These concerns are discussed, expanded upon, and illustrated with multiple cases in Edsall 1975.

15. See Jasanoff 1990, 34–35, for a quick overview of the multiple advisory panels created in the 1970s for the EPA, Food and Drug Administration (FDA), OSHA, and CPSC. Later chapters provide a more detailed look at the history of the advisory committees.

16. Examples of this type of controversy in the 1970s include the regulation of pesticides such as 2,4,5-T (Smith 1992, 78–82), dieldrin, aldrin, heptachlor, and chlordane (see Karch 1977); debates over air quality standards (Jasanoff 1990, 102–12); and the debate over saccharin.

17. For examples, see Jasanoff 1990, chap. 2, which discusses science advice on nitrites, 2,4,5-T, and Love Canal.

18. Instead, commentary about science advising has been the province of scien-

tists themselves and of social scientists, especially political scientists. See Price 1954, Wolfle 1959, Leiserson 1965, and Dupre and Lakoff 1962.

19. There were echoes of this ideal before the 1950s (such as Compton's call for "impartial advice" in 1936, quoted above), but what Compton and others mean by this value-free science advice is not clear. Further research into the history of these norms in the twentieth-century American scientific community is needed. The well-defined norm I will discuss in the next chapter has a clear locus of origin in the post–World War II context.

Chapter 3: Origins of the Value-Free Ideal for Science

1. Hence, there exists general neglect of research ethics in philosophy of science. A new journal, *Science and Engineering Ethics*, was begun in 1995 by scientists and engineers to fill the gap left by philosophers of science. About the same time, a few philosophers of science (such as Kristin Shrader-Frechette [1994] and David Resnick [1998]) published books on research ethics, but the topic has not been, nor since become, a central topic in philosophy of science.

2. Not all experts in technical disputes are honest participants, but many are. How to tell honest from dishonest ones will be discussed in chapters 5, 6, and 7.

3. As I note below, the value-free ideal for science has a complex history reaching back several centuries. However, what has been meant by that ideal has shifted over time. See Proctor 1991 for one account of those shifts. I will not recount that history here, but instead will focus on the history of the value-free ideal as it is currently formulated.

4. A welcome respite from this drought has been the work of Shrader-Frechette (1991, 1993, 1994) and the collection of essays in Mayo and Hollander (1991). However, I diverge from Shrader-Frechette's views on values in science, as noted in chapter 1.

5. The value-free ideal was held by less prominent philosophical perspectives, such as the neo-Thomists at the University of Chicago, who subscribed to a strong fact-value distinction and to a clear-cut separation of science and values. Their views were thought to be erroneous and juvenile by their pragmatist and logical empiricist critics (Reisch 2005, 72–80).

6. Merton introduced the ethos of science in his 1938 essay, "Science and the Social Order," first published in *Philosophy of Science* 5:321–37.

7. See Proctor 1991, 201–3, for more on Ayer's work. In the same chapter, Proctor suggests that Carnap held similar views, but recent scholarship has more sharply distinguished among the analytic philosophers of the period. See, for example, Richardson 2003 and Reisch 2005, 47–53, 83–95.

8. See Uebel 2003, Howard 2003, and Reisch 2005, 27–56, for more details on the views of Frank, Carnap, and Neurath.

9. See Reichenbach 1938, 6–7, for the premiere of the distinction. See Howard 2003, 53–54, for a brief discussion of Reichenbach in this period.

10. This appears to be what happened to Philip Frank's work. See Reisch 2005, 307–30.

11. See Schrecker 1986 for an account of academia in general during this period;

McCumber 2001 for an account of philosophy; and Reisch 2005 for an account of the philosophy of science.

12. In a letter from the Philosophy of Science Association president C. J. Ducasse to outgoing *Philosophy of Science* editor-in-chief C. West Churchman in September 1958, Ducasse elaborated on his desired direction for the journal:

> I think there should be room in the Journal not only for formalized articles, and for articles on methodology, but also for articles on such topics as, say, the naive tacit metaphysics of scientists who repudiate having any metaphysics; on the psychology of logicians and scientists in general; on the sociology of science; on such a question as that of the relation between philosophy of science and theory of knowledge; on the relation between proving scientifically, and convincing; and so on.
>
> The one thing I would insist on, however, in such articles no less than in formalized ones, or methodological or specialized ones, would be competence at least, and preferably, high quality—for such articles, as well as formalized ones, can have this. We should have not half-baked or amateurish stuff; and we should try to have some articles by men of high reputation fairly often. (Richard Rudner Archives, Washington University, St. Louis)

13. Indeed, Reichenbach's conception of knowledge as inherently value free was not the value-free ideal that would be accepted by the 1960s.

14. The centrality of this debate for philosophy of science at the time can be seen in the fact that Jeffrey thanked Carl Hempel, "at whose suggestion this paper was written" (Jeffrey 1956, n1). Hempel 1965 encapsulates much of the debate he helped to instigate.

15. In his own examples, the probabilities assigned are simply assumed to be reliable and "definite" (Jeffrey 1956, 241), and thus Jeffrey failed to grapple with Rudner's concern. The probabilistic framework of which Jeffrey wrote, the Bayesian framework, has yet to be shown useful in real world contexts where both likelihoods and priors are disputed. At best, it serves as general guidance for how one's views should shift in the face of evidence. When both likelihoods and priors are disputed, abundant evidence may still never produce a convergence of probability.

16. Note how this echoes Merton's view of the ethos of science (Merton 1938, 1942) while rejecting the social complexity in which Merton saw the ethos as being embedded.

17. Levi also criticized Rudner's arguments for supposedly relying on a simplistic view of the relationship between beliefs and actions, a simple behavioralism wherein the acceptance of a belief automatically led to some action. However, as Leach 1968 and Gaa 1977 demonstrate, Rudner requires no such behavioralism for his arguments to work, and Rudner made no claims for such a behavioralism. Rudner needs only the rather obvious understanding that beliefs are used to make decisions about actions, and thus have a clear influence on actions.

18. Although many philosophers have contributed to the philosophy of science since ancient times, philosophy of science as its own discipline did not emerge until the mid-twentieth century. Documenting this development is the task for another paper.

19. Oddly, Leach (1968) cites Nagel's work as being a defender of the value-neutrality thesis. A careful reading of Nagel reveals more ambivalence than in Jeffrey 1956 or Levi 1960 and 1962, however.

20. See Longino 1990, chap. 4, and Longino 2002, chaps. 5 and 6.

21. The essay was first published in R. E. Spiller, ed. (1960), *Social control in a free society*, Philadelphia: University of Pennsylvania Press, and later reprinted in *Aspects of scientific explanation* (Hempel 1965a) with little change.

22. Hempel lays out the distinction between instrumental and categorical value judgments. According to Hempel, science is useful for the instrumental value judgments, but not categorical value judgments. In his view, science can inform our choices by telling us how to achieve our goals. It can even influence our ideas about which goals are achievable and thus which are worth pursuing intently. Science cannot tell us ultimately what we *should* value, however. Frank and others may have disagreed with Hempel on this point, but it is an issue beyond the scope of this work.

23. Unlike Jeffrey, Hempel sees the acceptance or rejection of hypotheses as an important part of science.

24. This book appeared in 1962 as the final installment in the Foundations of the Unity of Science series (a series led by Carnap, Frank, and Otto Neurath, before Neurath's death in 1945). See Reisch 2005, 229–33, for a discussion of Kuhn and Frank, and the reception of their work in the late cold war. Reisch suggests that Kuhn viewed science as a discipline to be isolated from philosophy as well as the public. For a sweeping account of Kuhn and his influence, see Fuller 2000.

25. Laudan 1984 focuses solely on the so-called "cognitive" aims of science:

> In this book I talk quite a lot about values, and to that extent the title is apt; but I have nothing to say about ethical values as such, for they are manifestly not the predominant values in the scientific enterprise. Not that ethics plays no role in science; on the contrary, ethical values are always present in scientific decision making and, very occasionally, their influence is of great importance. But that importance fades into insignificance when compared with the ubiquitous role of cognitive values. (xii)

There is no discussion or acknowledgment that any other aims of science (other than those relating to truth in some way) should be considered to understand science; they are "insignificant." Although Laudan does not exclude them from relevance completely, he is concerned with an internal understanding of scientific change, internal to scientists and their disputes isolated from society at large, and with understanding how scientists appear to find consensus among themselves.

26. The closest thing to a direct critique of Kuhn's isolationist view of science comes from Feyerabend (1964), who took issue with Kuhn's endorsement of scientific dogmatism. Feyerabend was responding to an essay by Kuhn published in a collection on scientific change edited by A. C. Crombie. Dogmatism is the result of adherence to a paradigm for Kuhn, and the sign of a mature science. Dogmatism produces a focus on the puzzles a paradigm generates rather than other problems science might address and solve, problems that would come from outside of disciplinary science.

27. Leach 1968 is a clear attack on the value-free ideal, on the same grounds as Rudner had argued against it in 1953. Leach attempts to reintroduce the pragmatic

concerns of Rudner and Churchman, noting "that the amount of evidence necessary to warrant the acceptance of any particular explanatory hypothesis is not fixed but instead depends on our varied purposes," including practical as well as epistemic purposes (93). He also presents careful arguments against Levi's critique of Rudner. The essay drew little attention by philosophers of science. Philosophers also neglected Scriven 1974.

28. In a web of science citation search, Gaa's essay was cited only in three economics articles, all by the same author, and by no philosophy of science work.

29. McMullin (2000, 553) still sees applications of science as distinct from science, and thus the role of values in science is disconnected from the public uses of science.

30. As noted in chapter 1, and discussed further in chapter 5, there have been challenges to the value-free ideal, particularly coming from the work of feminist philosophers of science. Nevertheless, the value-free ideal is still widely held and defended.

31. Lacey labels the by now traditional value-free ideal "impartiality" in his book to distinguish it from other stronger theses regarding the role of values in science.

32. Because of this limited definition Lacey 1999 seriously misconstrues Rudner's 1953 arguments. See Lacey 1999, 71–74.

Chapter 4: The Moral Responsibilities of Scientists

Some ideas and passages in this chapter are drawn from H. Douglas, "The moral responsibilities of scientists: Tensions between autonomy and responsibility," *American Philosophical Quarterly* 40:59–68, © 2003 by North American Philosophical Publications, Inc., with kind permission of North American Philosophical Publications, Inc., and from H. Douglas, "Rejecting the ideal of value-free science," in *Value-free science?* edited by Kincaid, Dupre, and Wylie (2007), 120–38, with kind permission of Oxford University Press, Inc.

1. The issue of whether social and ethical values can in fact be clearly demarcated from cognitive or epistemic values will be addressed in the next chapter. For now, I will assume that they can be. Of course, if they cannot be, and the demarcation of values the current value-free ideal rests upon fails, then the ideal is untenable. This chapter addresses instead whether it is desirable.

2. For these kinds of discussions, see, for example, Arnold 2001, Fischer and Ravizza 1993, Fischer 1999, and Paul, Miller, and Paul 1999.

3. We are held morally responsible for both choice of actions and the consequences of actions. These two can be distinct. A person may be held morally responsible for a bad choice (or a good one), even if no significant consequences follow from their choice. However, if there is no "choice" involved (for example, the person was sleepwalking or under strong coercion), we generally do not hold the person responsible for the consequences of their actions. Thus, we can evaluate a choice irrespective of the actual consequences of the example.

4. See Hardimon 1994, esp. 337–42, for a discussion of the idea that one's role obligations suffice for fulfilling all moral obligations (the doctrine of perfect adequacy) and for reasons why Hardimon rejects this doctrine, as do most contemporary moral theorists.

5. Unfortunately, little detailed examination or discussion of a moral exemption for scientists has occurred in the past fifty years. See Gaa 1977 for a similar diagnosis.

6. See Greenberg 1967, 171–208, for a detailed account of Mohole's rise and fall.

7. H. G. Wells had speculated in *The world set free* on the topic before World War I, but as there were no known processes for making his speculation a reality, it remained in the realm of science fiction.

8. See Rhodes 1986, 233–75, for a clear account of this story.

9. The general foreseeability of the bomb after the discovery of fission is underscored by the surprising fact that Einstein did not think of the implications of fission until Leo Szilard explained them to him in the summer of 1939. It would not be a surprise if it were not generally foreseen elsewhere.

10. This judgment always contains the combination of the judgment of the size of the uncertainty, the judgment of what the potential consequences of error are, and the importance of those consequences.

Chapter 5: The Structure of Values in Science

1. See Kuhn 1977 for a more detailed elaboration on these values. In addition, Kuhn argues that these epistemic values are best interpreted as *values* because scientists weight and interpret them differently, thus producing variable theory choice among scientists, even in the face of the same body of evidence.

2. In the years since 1960, the division between the values acceptable in science and those not acceptable in science has been described in multiple ways. The traditional epistemic versus nonepistemic dichotomy has been augmented by a cognitive versus noncognitive distinction, and a constitutive versus contextual contrast. Any one of these terms—epistemic, cognitive, constitutive—is often used to describe values at the heart of Levi's canon, and is used to contrast those values to be excluded, although in the case of Longino's (1990) contextual/constitutive distinction, no such normative force lies behind it.

3. Machamer and Douglas 1998 and 1999 also emphasize the untenability of the epistemic versus nonepistemic value distinction.

4. I will not make any attempt to determine answers to the difficult ethical questions that lurk under this arena. For example, I will not argue for or against the use of economic assessments of harm in a cost-benefit framework to determine whether a risk is ethically acceptable. I will lump all of these considerations into the realm of ethical values, which are our guide for deciding what the right thing to do is. Actually weighing ethical values is a task for another body of work.

5. Although "precision" and "accuracy" have specific meanings in certain scientific disciplines, I do not mean to invoke them here. Rather, by "precision," I mean something akin to the number of decimals one is able to assign to a measurement (or a prediction), while "accuracy" means how close one gets to the target (either the prediction or the actual measurement).

6. Fraassen (1980) argues that acceptance of a scientific theory is tantamount to accepting that it is empirically adequate, that is, that it maps well with our currently observable universe (Fraassen 1980, 12, 45–46). This move was made as part of the realist debate, but it has since been considered a minimalist criterion for scientific work.

7. To use Longino's most recent terminology; see Longino 2002, 115–18.

8. Epistemic criteria provide necessary conditions for science, but not necessary and sufficient. They are not meant to solve the demarcation problem of defining what counts as science as opposed to everything else.

9. This restriction on a direct role for values captures a key aspect of the distinction between theoretical and practical reason, in my view, an aspect that retains its importance even for the *action* of making an empirical claim.

10. Hempel 1965b and Rudner 1953 both acknowledge this role for values in science. Philosophers of science have begun to pay increased attention to the role for values in the selection of scientific research projects, delving more deeply into how we should think about which science to pursue. Lacey 1999 and Kitcher 2001 have laudably begun this discussion, but much more work needs to be done.

11. We have rightly condemned the Nazi's experiments on helpless human beings, but, sadly, similar episodes mar experimental practice in the United States as well. The infamous Tuskegee syphilis experiments, in which African-American men with syphilis went deliberately and deceptively untreated, continued until the early 1970s. See Bulger 2002 for a concise overview of the history and issues.

12. There may also be some specific knowledge that we would prefer not to have, such as how to more easily accomplish isotope separation for heavy elements (see Sinsheimer 1979, 29). This issue touches on the topic of "forbidden knowledge" and whether there are pieces of knowledge we would be better off without. A detailed examination of this issue is beyond the scope of this work. But the decision to not pursue a project for such reasons would fit into the schema here for values in science, being a direct role for values in the decision of which projects to pursue. However, it would be a direct role for values that would undermine the value we place in developing new knowledge, the social value that underlies the scientific endeavor. Hence the great tension around the topic of forbidden knowledge.

13. This simplified case is inspired by Longino's more nuanced accounts in Longino 1990, chaps. 6 and 7.

14. See Graham 1987, 102–50, for a richer account of the Lysenko affair.

15. See Drake 1967 for a brief history of Galileo and the church.

16. Of course, our values direct our actions, which end up shaping our world in important and substantial ways. But this kind of influence has no direct relevance to the empirical claims we should make right now.

17. Much of the above paragraph is drawn from H. Douglas, "Inductive risk and values in science," *Philosophy of Science* 67:559–79, © 2000 by the Philosophy of Science Association, with kind permission of the Philosophy of Science Association.

18. Traditionally in science in the twentieth century, when tests of statistical significance became widespread, it was thought that avoiding false positives was more important than avoiding false negatives. The reasoning seems to have been that it was very important to keep a false statement from being accepted into the canons of scientific thinking, much more so than to reject a true statement incorrectly. Yet this line of thinking was developed before science took up its prominent place in public life. It is doubtful that focusing solely on the concerns internal to scientific work are an acceptable guide here, given that scientists who depend on the work of other scientists often rely upon the data themselves rather than a final interpretation and given the

requirements of moral responsibility. Cranor 1993 argues that, in fact, false negatives cost society far more than false positives and thus that the current levels are out of balance.

19. As I will discuss in the next chapter, such value-neutrality may have its uses in some contexts. But it is not always the best or even a desirable choice.

20. Perhaps evidence could be gathered to show that in fact these cognitive values do increase the likelihood a theory is true. Such evidence has not yet been gleaned from the history of science. Instead, cases on both sides have been constructed, usually with pro-cognitive value theorists pointing to still accepted theories and showing how they instantiate cognitive values and skeptics finding counterexamples in history where theories instantiated such values and were ultimately rejected. Perhaps a careful statistical analysis of each of these values would tell us more definitively whether they have epistemic import, but given how difficult it is to decide what counts as a theory reflecting a particular value, the difficulties of cleanly delineating the values, and the need for a complete count of theories in the history of science (another theoretical nightmare), it is doubtful that such an analysis could be performed and be acceptable to science scholars.

21. I draw this history from Dutton 1988, chap. 3, which provides a clear and informative history of DES and its use in the United States. Dutton is primarily focused on issues of government oversight and the ethical treatment of patients. I will focus on the reasoning among the scientists who made claims about DES. My examination here is admittedly brief and glosses many details; a fuller treatment will be forthcoming.

22. Other scientists were either more cautious, noting the potential problems with estrogenic compounds, or had more sophisticated models of how DES would be helpful.

Chapter 6: Objectivity in Science

The ideas and some passages in this chapter are drawn from H. Douglas, "The irreducible complexity of objectivity," *Synthese* 138:453–73, © 2004 by Kluwer Academic Publishers, with the kind permission of Springer.

1. If one wants to assess whether a person or a process is objective, examining the attributes I describe below would be helpful. However, I do not want to argue that objectivity is first and foremost about knowledge claims, and the objectivity of persons and processes derivative.

2. Traditionally, an objective result would be one that gained a grasp of the real objects in the world. However, whether or not it is possible to have good evidence that one has gained such a grasp is precisely at the heart of the realist debates, debates that seem to be solidifying into convictions on each side that their side has won. The view on objectivity I articulate here attempts to remain agnostic over the realism issue. This view, I hope, can be drawn on by either side to articulate what is meant by objectivity, although realists will want to make some additional claims about the implications of having objectivity present. In particular, a realist will want to claim that the presence of the process markers I describe below means that one can make strong claims about the knowledge product. The realist, on the one hand, will make the leap from the process markers of objectivity to the conclusion that we have grasped real objects. The antirealist, on the other hand, can acknowledge the importance of these process mark-

ers as indicative of reliable knowledge while denying the realist conclusion (or weakening it in some way). While both camps may agree that the processes produce results they would label "objective," one should not then assume that the objective product (whatever is produced by the proper process) has additional "objective" characteristics. I call the product "objective" solely because the process of its production has certain markers. Additional (realist) arguments are needed to move from the bare bones objectivity arising from an examination of process to the ontological claims about the product, that the objects described are "really there."

3. Even if one is a realist, one must be careful not to inflate the claims made in the name of this aspect of objectivity. Even if one has the strongest sense possible that one has gotten at an object, that some object really is there and that one can manipulate it in some reliable way, this does not mean that all of the theories about that object are true, or that we know everything there is to know about it. All we know is that there is something really *there* to work with. Biologists that can reliably manipulate a cellular receptor do not doubt that it exists, even if they do not fully understand its function (as with, for example, the Ah receptor), and chemists do not doubt that chemicals commonly used to perform reactions exist, even if all of the mysteries of the chemical are not fully plumbed. This lack of doubt among scientists does not exclude the possibility of surprise (for example, that the object will fracture into something more complex) or error (for example, that two different objects will come to be seen as the same entity under different circumstances). Broad claims of realism (or closeness to truth) for scientific theories *as a whole* are not supportable by manipulable objectivity.

4. Kosso 1989 focuses on this aspect of objectivity.

5. This is also the primary sense of objectivity highlighted by those concerned with "invariance" in experience (although in some cases manipulable objectivity may also be in play) (see Nozick 1998). Unfortunately, Nozick's argument that invariance captures all of objectivity falls short. First, the additional aspects of objectivity I discuss below are merely noted as assisting with the gaining of invariance (even though this is not always the case) and thus Nozick fails to capture their importance. Second, invariance depends on comparisons across approaches, but scientists must decide which approaches are acceptable. This may reduce in practice to scientists telling us what is objective and what is not. Finally, it is not clear whether Nozick's invariance captures objectivity or theoretical sophistication. Invariance may just mean that theoretical development has advanced to the point that one can convert an outcome from one system or perspective to another, not that measurements actually taken capture something of the world. For example, one can have invariance between Celsius and Fahrenheit temperature measurements (because there is a set way to convert from one scale to the other), but that does not show that thermometers are capturing something reliably. To show that, it would be better to use a more independent check on temperature, such as an infrared scan. If the scan produces a temperature that is close to the thermometer's measure (that is, is convergent with it), one gets the sense that the measurements are reliable and objective. The match between the two measures need not be exact, contrary to Nozick's sense of invariance.

6. See Kosso 1989 for a detailed examination of epistemic independence.

7. As in the recent Science Advisory Board review of the EPA's dioxin health assessment revisions, where the SAB panel found that "the Agency document contains a

quite thorough and generally objective summarization of that [peer-reviewed dioxin] literature" (EPA/SAB 2001, 16).

8. See Megill 1994, 10–11. Note that Megill's definition of procedural objectivity has far more in common with Porter's (1995, 4) and Daston and Gallison's (1992, 82–83) use of "mechanical" objectivity than with Fine's (1998, 11) use of the label. Fine's definition of "procedural" objectivity seems to include all three senses of objectivity concerned with social processes that I discuss here.

9. Porter 1995 labels this sense of objectivity "mechanical" instead of "procedural." I use the latter term because it seems to capture better this sense of objectivity.

10. In practice, there can be strong ties between the three senses of objectivity related to social processes. Agreed upon procedures for making observations (procedural objectivity) can greatly promote the achieving of concordant objectivity. The final sense of objectivity to be introduced below, interactive objectivity, can help define the disciplinary boundaries in which such agreements on procedures (and then observations) take place. However, while such interdependence can occur, it need not. The senses are still conceptually distinct and can be independent of one another.

11. See Harrison 2005 for a moving firsthand account of this sighting.

12. This is not to say that all skeptics were convinced. But the objectivity of the claim was sufficient to get $10 million from the U.S. government for further research and habitat protection (Dalton 2005).

13. See Jones and Wigley 1990 for an overview of the traditional temperature-taking methods.

14. Some critics have accused the IPCC of not being open to full discussion. For an account of such criticism and a rebuttal, see Edwards and Schneider 2001.

Chapter 7: The Integrity of Science in the Policy Process

1. See Fiorino 1995, 23–32, for an overview of the environmental laws listed here.

2. To fulfill its statutory obligations in an age when absolute safety cannot be guaranteed, the EPA has decided that a lifetime cancer risk of 1 in 10,000 is an acceptable risk. They then produce an "ample margin of safety" for this acceptable risk by regulating to reduce the risk to 1 in a million (NRC 1994, 3).

3. A similar point can be found in Sarewitz 1996, chap. 5.

4. Lowrance (1976) wrote his monograph at the behest of the National Academy of Science's Committee on Science and Public Policy.

5. The NRC argued for conceptual and procedural separation between risk assessment and risk management, but not an institutional separation. If one agency performed risk assessments and another risk management, the NRC warned that the risk assessments produced would likely be useless to actual policymaking.

6. William Ruckelhaus, who headed the EPA in the mid-1980s and attempted to institute the NRC's recommendations concerning risk assessment, presented two different reasons to keep risk assessment separate from risk management. The first reason, which he articulated in 1983, was to keep the science in risk assessment from being adversely affected by political issues (Ruckelhaus 1983). The second reason, articulated two years later, was to keep the technical experts from taking over the "democratic" risk management process (Ruckelhaus 1985). The first reason seems to have had more historical importance, particularly in light of the scandals arising in

the Reagan-Gorsuch EPA (1981–83) and the administration's deregulation policies (see Silbergeld 1991, 102–3).

7. In addition, it was hoped that the development and use of guidelines would spur the development of the new field of risk assessment, including research into the basic issues underlying risk assessment (NRC 1983, 72).

8. The attempts to develop inference guidelines have been rife with controversy since the 1970s. See Karch 1977; Landy et al. 1990, chap. 6, for detailed accounts of these attempts; and NRC 1994, chap. 6 for commentary on the difficulties.

9. The first official dioxin risk assessment was completed in 1986. In 1991, the EPA began the process of drafting a new risk assessment. It was the most recent draft of this reassessment the SAB was asked to examine.

Chapter 8: Values and Practices

1. My use of the terms "deliberative" and "participatory" should not invoke the debates occurring in political theory concerning deliberative and participatory democracy. In those discussions, "deliberative democracy" refers to ideals of public discourse across the broadest scale of issues and political spectra, and "participatory democracy" refers to sectors of the public being engaged in political activism, particularly through parties or social movements and protests. This kind of participation is not what I mean to invoke here, and indeed, it is hard to see how such a form of participation could be helpful in risk assessments or particular science policy decisions. In addition, the ideals for deliberation in deliberative democracy are usually examined with respect to creating more casual political discourse in everyday settings, again settings less relevant to the context of science policy. Mutz 2006 argues that tensions exist between activist participation and everyday casual deliberations (or sharing of reasons) across political lines. While she is surely correct in this observation, the situations for deliberation concerning risk and values are not casual but must be structured, and indeed structured so as to include more than just activists.

2. See Laird 1993 and Fischer 1993 for more details on the benefits of public participation, particularly as a transformative process. Many attempts to develop public participation are reminiscent of Dewey 1927, where he called for experimentation with processes that involve the public more with governance.

3. Again, "deliberation" is taken here to mean simply an exchange and consideration of reasons, something required for interactive objectivity, and should not invoke the full theoretical trappings of deliberative democracy theory.

4. Passages from the preceding three paragraphs are drawn from H. Douglas, "Inserting the public into science," in *Democratization of expertise? Exploring novel forms of scientific advice in political decision-making*, edited by Maasen and Weingart (2005), 159–60, with the kind permission of Springer.

5. There is a growing body of literature on public participation in policy processes. NRC 1996, appendix A, 167–98, has six examples of some public involvement with risk analysis. See also Fischer 1993 and 2000, Futrell 2003, Irwin 1995, Laird 1993, Lynn and Busenberg 1995, Renn 1999, and Rowe and Frewer 2000 for various accounts. Douglas 2005 provides another discussion of some of these examples and how to evaluate the promise of analytic-deliberative processes.

6. See Boiko et al. 1996 for an examination of different processes to identify members of the public for involvement.

7. Again, I will focus here on value deliberations centered on particular policy issues rather than deliberative democracy's general goal of improving public political discourse across the full range of lived contexts and political issues.

8. Indeed, the NRC (1996, 98) suggested that some use of ethics, in particular "the application of principles of ethics," could be considered among the qualitative methods of analysis. I am skeptical that there are principles of sufficient agreement to meet the requirements for analysis, but the thought experiment approach I describe here might meet such requirement.

9. As Kant does in *Foundations of the metaphysics of morals* (Kant 1785).

10. This is similar to the original position found in Rawls 1971.

11. A similar concern is raised in Kitcher 2001, chap. 10, although it is focused there on the distribution of funds for scientific research, or policy for science.

REFERENCES

Arnold, Denis. 2001. Coercion and moral responsibility. *American Philosophical Quarterly* 38:53–67.
Barnes, Barry, and David Bloor. 1982. Relativism, rationalism, and the sociology of knowledge. In *Rationality and relativism*, ed. Martin Hollis and Steven Lukes, 21–47. Cambridge, MA: MIT Press.
Bazelon, David. 1979. Risk and responsibility. *Science* 205:277–80.
Berger, Bennett. 1994. Taking arms. *Science* 264:985–89.
Boghossian, Paul. 1998. What the Sokal Hoax ought to teach us. In *A house built on sand: Exposing postmodernist myths about science*, ed. Noretta Koertge, 23–31. New York: Oxford University Press.
Boiko, Patricia, Richard Morrill, James Flynn, Elaine Faustman, Gerald van Belle, and Gilbert Omenn. 1996. Who holds the stakes? A case study of stakeholder identification at two nuclear weapons production sites. *Risk Analysis* 16:237–49.
Breyer, Stephen. 1998. The interdependence of science and law. *Science* 280:537–38.
Bridgman, P. W. 1947. Scientists and social responsibility. *Scientific Monthly* 65:48–154.
Broad, William, and Nicholas Wade. 1982. *Betrayers of the truth*. New York: Simon and Schuster.
Bronk, Detlev W. 1974. Science advice in the White House. *Science* 186:116–21.
———. 1975. The National Science Foundation: Origins, hopes, and aspirations. *Science* 188:409–14.
Brooks, Harvey. 1964. The scientific advisor. In *Scientists and national policy-making*, ed. Robert Gilpin and Christopher Wright, 73–96. New York: Columbia University Press.
———. 1968. Physics and the polity. *Science* 160:396–400.
———. 1975. Expertise and politics—Problems and tensions. *Proceedings of the American Philosophical Society* 119:257–61.
Bulger, Ruth Ellen. 2002. Research with human beings. In *The ethical dimensions of the biological and health sciences*, ed. Ruth Ellen Bulger, Elizabeth Heitman, and Stanley Joel Reiser, 177–25. New York: Cambridge University Press.

Burger, Edward. 1980. *Science at the White House: A political liability*. Baltimore: Johns Hopkins University Press.

Busenberg, George. 1999. Collaborative and adversarial analysis in environmental policy. *Policy Sciences* 32:1–11.

Bush, Vannevar. 1945/1960. *Science: The endless frontier*. Repr., Washington, DC: National Science Foundation.

Callahan, Joan. 1994. Professions, institutions, and moral risk. In *Professional ethics and social responsibility*, ed. Daniel E. Wueste, 243–70. Lanham, MD: Rowman and Littlefield.

Chariff, Russel A., Kathryn A. Cortopassi, Harold K. Figueroa, John W. Fitzpatrick, Kurt M. Fristrup, Martjan Lammertink, M. David Luneau Jr., Michael E. Powers, and Kenneth V. Rosenberg. 2005. Notes and double knocks from Arkansas. *Science* 309:1489.

Churchman, C. West. 1948a. Statistics, pragmatics, induction. *Philosophy of Science* 15:249–68.

———. 1948b. *Theory of experimental inference*. New York: The MacMillan Company.

———. 1956. Science and decision making. *Philosophy of Science* 22:247–49.

Compton, Karl T. 1936. Science advisory service to the government. *Scientific Monthly* 42:30–39.

Cranor, Carl F. 1993. *Regulating toxic substances: A philosophy of science and the law*. New York: Oxford University Press.

Culliton, Barbara. 1979. Science's restive public. In *Limits of scientific inquiry*, ed. Gerald Holton and Robert S. Morrison, 147–56. New York: W. W. Norton.

Dalton, Rex. 2005. A wing and a prayer. *Nature* 437:188–90.

Daston, Lorraine. 1992. Objectivity and the escape from perspective. *Social Studies of Science* 22:597–618.

Daston, Lorraine, and Peter Gallison. 1992. The image of objectivity. *Representations* 40:81–128.

Dewey, John. 1927. *The public and its problems*. Athens, OH: Ohio University Press.

Dieckmann, W. J., M. E. Davis, L. M. Rynkiewicz, and R. E. Pottinger. 1953. Does the administration of diethylstilbestrol during pregnancy have therapeutic value? *American Journal of Obstetrics and Gynecology* 66:1062–75.

Douglas, Heather. 1998. *The use of science in policy-making: A study of values in dioxin science*. Ph.D. diss., University of Pittsburgh.

———. 2000. Inductive risk and values in science. *Philosophy of Science* 67:559–79.

———. 2003a. Hempelian insights for feminism. In *Siblings under the skin: Feminism, social justice, and analytic philosophy*, ed. Sharyn Clough, 238–306. Aurora, CO: Davies Publishing Group.

———. 2003b. The moral responsibilities of scientists: Tensions between autonomy and responsibility. *American Philosophical Quarterly* 40:59–68.

———. 2004. The irreducible complexity of objectivity. *Synthese* 138:453–73.

———. 2005. Inserting the public into science. In *Democratization of expertise? Exploring novel forms of scientific advice in political decision-making*, ed. Sabine Maasen and Peter Weingart, 153–69. Dordrecht: Springer.

———. 2006. Bullshit at the interface of science and policy: Global warming, toxic substances, and other pesky problems. In *Bullshit and philosophy*, ed. Gary L. Hardcastle and George A. Reisch, 213–26. La Salle, IL: Open Court.

———. 2007. Rejecting the ideal of value-free science. In *Value-free science? Ideals and illusions*, ed. Harold Kincaid, John Dupré, and Alison Wylie, 120–39. New York: Oxford University Press.

Drake, Stillman. 1967. Translator's preface to *Dialogue concerning the two chief world systems*, by Galileo Galilei, xxi–xxvii. Berkeley and Los Angeles: University of California Press.

Dupre, J. Stefan, and Sanford A. Lakoff. 1962. *Science and the nation: Policy and politics*. Englewood Cliffs, NJ: Prentice-Hall.

Dupree, Hunter. 1957. *Science in the federal government: A history of policies and activities to 1940*. Cambridge, MA: Harvard University Press.

Dutton, Diana. 1988. *Worse than the disease: Pitfalls of medical progress*. New York: Cambridge University Press.

Edsall, John T. 1975. *Scientific freedom and responsibility: A report of the AAAS Committee on Scientific Freedom and Responsibility*. Washington, DC: American Association for the Advancement of Science.

Edwards, Paul N., and Stephen H. Schneider. 2001. Self-governance and peer review in science-for-policy: The case of the IPCC second assessment report. In *Changing the atmosphere: Expert knowledge and environmental governance*, ed. Clark A. Miller and Paul N. Edwards, 219–46. Cambridge, MA: MIT Press.

Environmental Protection Agency. Scientific Advisory Board. 2001. *Dioxin reassessment—An SAB review of the Office of Research and Development's reassessment of dioxin*. EPA-SAB-EC-01-006.

———. 2005. *Guidelines for carcinogen risk assessment*. EPA/603/P-03/001B.

Fancher, Raymond. 1985. *The intelligence men: Makers of the IQ controversy*. New York: W. W. Norton.

Fausto-Sterling, Anne. 1985. *Myths of gender: Biological theories about women and men*. New York: Basic Books.

Feinberg, Joel. 1970. *Doing and deserving*. Princeton, NJ: Princeton University Press.

Feyerabend, Paul. 1964. Review of *Scientific change*, by A. C. Crombie. *British Journal of Philosophy of Science* 15:244–54.

Fine, Arthur. 1998. The viewpoint of no-one in particular. *Proceedings and Addresses of the APA* 72:9–20.

Fiorino, Daniel. 1990. Citizen participation and environmental risk: A survey of institutional mechanisms. *Science, Technology, and Human Values* 15:226–43.

———. 1995. *Making environmental policy*. Berkeley and Los Angeles: University of California Press.

Fischer, Frank. 1993. Citizen participation and the democratization of policy expertise: From theoretical inquiry to practical cases. *Policy Sciences* 26:165–87.

———. 2000. *Citizens, experts, and the environment: The politics of local knowledge*. Durham, NC: Duke University Press.

Fischer, John Martin. 1999. Recent work on moral responsibility. *Ethics* 110:93–139.

Fischer, John Martin, and Mark Ravizza, eds. 1993. *Perspectives on moral responsibility.* Ithaca, NY: Cornell University Press.

Fitzpatrick, John W., Martjan Lammertink, M. David Luneau Jr., Tim W. Gallagher, Bobby R. Harrison, Gene M. Sparling, Kenneth V. Rosenberg, et al. 2005. Ivory-billed woodpecker (*Campephilus principalis*) persists in continental North America. *Science* 308:1460–62.

Flory, James, and Philip Kitcher. 2004. Global health and the scientific research agenda. *Philosophy and Public Affairs* 32:36–65.

Fraassen, Bas van. 1980. *The scientific image.* Oxford: Oxford University Press.

Frank, Philipp G. 1953. The variety of reasons for the acceptance of scientific theories. In *The validation of scientific theories*, ed. Philipp G. Frank, 13–26. New York: Collier Books.

Fuller, Steve. 2000. *Thomas Kuhn: A philosophical history for our times.* Chicago: University of Chicago Press.

Futrell, Robert. 2003. Technical adversarialism and participatory collaboration in the U.S. chemical weapons disposal program. *Science, Technology, and Human Values* 28:451–82.

Gaa, James. 1977. Moral autonomy and the rationality of science. *Philosophy of Science* 44:513–41.

Gilpin, Robert. 1964. Introduction: Natural scientists in policy-making. In *Scientists and national policy-making*, ed. Robert Gilpin and Christopher Wright, 1–18. New York: Columbia University Press.

Gilpin, Robert, and Christopher Wright, eds. 1964. *Scientists and national policy-making.* New York: Columbia University Press.

Gough, Michael, ed. 2003. *Politicizing science: The alchemy of policymaking.* Washington, DC: Hoover Institution Press.

Gould, Stephen Jay. 1981. *The mismeasure of man.* New York: W. W. Norton.

Grafton, Anthony. 1990. *Forgers and critics: Creativity and duplicity in Western scholarship.* Princeton, NJ: Princeton University Press.

Graham, Loren. 1987. *Science, philosophy, and human behavior in the Soviet Union.* New York: Columbia University Press.

Greenberg, Daniel. 1999. *The politics of pure science.* Chicago: University of Chicago Press.

———. 2001. *Science, money, and politics: Political triumph and ethical erosion.* Chicago: University of Chicago Press.

Gross, Paul, and Norman Levitt. 1994. *Higher superstition: The academic left and its quarrels with science.* Baltimore: Johns Hopkins University Press.

Guston, David. 1999. Evaluating the first U.S. consensus conference: The impact of the citizen's panel on telecommunications and the future of democracy. *Science, Technology, and Human Values* 24:451–82.

———. 2000. *Between politics and science: Assuring the integrity and productivity of research.* New York: Cambridge University Press.

Guston, David H., and Kenneth Keniston, eds. 1994. *The fragile contract: University science and the federal government.* Cambridge, MA: MIT Press.

Hacking, Ian. 1983. *Representing and intervening*. New York: Cambridge University Press.

Hard, G. C. 1995. Species comparison of the content and composition of urinary proteins. *Food and Chemical Toxicology* 33:731–46.

Hardimon, Michael. 1994. Role obligations. *Journal of Philosophy* 91:333–63.

Harding, Sandra. 1986. *The science question in feminism*. Ithaca, NY: Cornell University Press.

———. 1991. *Whose science? Whose knowledge?* Ithaca, NY: Cornell University Press.

Harrison, Bobby. 2005. Phantom of the bayou. *Natural History* 114:18.

Hart, Roger. 1996. The flight from reason: Higher superstition and the refutation of science studies. In *Science wars*, ed. Andrew Ross, 259–92. Durham, NC: Duke University Press.

Heil, John. 1983. Believing what one ought. *Journal of Philosophy* 80:752–65.

———. 1992. Believing reasonably. *Nous* 26:47–82.

Hempel, Carl G. 1965a. *Aspects of scientific explanation*. New York: The Free Press.

———. 1965b. Science and human values. In *Aspects of scientific explanation*, 81–96. New York: The Free Press.

———. 1981. Turns in the evolution of the problem of induction. *Synthese* 46. Reprinted in *The philosophy of Carl G. Hempel: Studies in science, explanation, and rationality*, ed. James H. Fetzer, 344–56. New York: Oxford University Press, 2001.

———. 2001. *The philosophy of Carl G. Hempel: Studies in science, explanation, and rationality*. Ed. James H. Fetzer. New York: Oxford University Press.

Herken, Gregg. 2000. *Cardinal choices: Presidential science advising from the atomic bomb to SDI*. Stanford, CA: Stanford University Press.

Herrick, Charles, and Dale Jamieson. 2001. Junk science and environmental policy: Obscuring the debate with misleading discourse. *Philosophy and Public Policy Quarterly* 21:11–16.

Hewlett, Richard, and Oscar Anderson. 1962/1990. *The new world: A history of the United States Atomic Energy Commission*, vol. 1, 1939–1946. Repr., Berkeley and Los Angeles: University of California Press.

Hewlett, Richard, and Francis Duncan. 1962/1990. *Atomic shield: A history of the United States Atomic Energy Commission*, vol. 2, 1947–1952. Repr., Berkeley and Los Angeles: University of California Press.

Hippel, Frank von, and Joel Primack. 1972. Public interest science. *Science* 177:1166–71.

Hitchens, Christopher. 1998. Afterword. *Critical Quarterly* (London) 40:44–45.

Hoffmann, Roald. 2003. Why buy that theory? *American Scientist* 91:9–11.

Holton, Gerald. 1978. Sub-electrons, presuppositions, and the Millikan-Ehrenhaft dispute. *Historical Studies in the Physical Sciences* 9:161–224.

Howard, Don. 2003. Two left turns make a right: On the curious political career of North American philosophy of science at mid-century. In *Logical empiricism in North America*, ed. Alan Richardson and Gary Hardcastle, 25–93. Minneapolis: University of Minnesota Press.

Huber, Peter. 1991. *Galileo's revenge: Junk science in the courtroom*. New York: Basic Books.

Hull, David. 1988. *Science as a process*. Chicago: University of Chicago Press.
Irwin, A. 1995. *Citizen science: A study of people, expertise, and sustainable development*. London: Routledge.
Jasanoff, Sheila. 1990. *The fifth branch: Science advisors as policy-makers*. Cambridge, MA: Harvard University Press.
———. 1992. Science, politics, and the renegotiation of expertise at EPA. *Osiris* 7:195–217.
Jeffrey, Richard. 1956. Valuation and acceptance of scientific hypotheses. *Philosophy of Science* 22:237–46.
Johnson, Deborah. 1996. Forbidden knowledge and science as professional activity. *Monist* 79:197–217.
———. 1999. Reframing the question of forbidden knowledge in modern science. *Science and Engineering Ethics* 5:445–61.
Jones, Philip D., and Tom M. L. Wigley. 1990. Global warming trends. *Scientific American* (August): 84–91.
Joss, Simon, and John Durant, eds. 1995. *Public participation in science: The role of consensus conferences in Europe*. London: Science Museum.
Kant, Immanuel. 1785. *Foundations of the metaphysics of morals*. In *Kant, Selections*, trans. Lewis White Beck. New York: Macmillan, 1988.
Karch, Nathan. 1977. Explicit criteria and principles for identifying carcinogens: A focus of controversy at the Environmental Protection Agency. In *Decision making in the Environmental Protection Agency*, vol. 2a, *Case Studies*, 119–206. Washington, DC: National Academy Press.
Kelly, Thomas. 2002. The rationality of belief and some other propositional attitudes. *Philosophical Studies* 110:163–96.
Kerr, Richard. 2004. Getting warmer, however you measure it. *Science* 304:805–06.
Kevles, Daniel. 1995. *The physicists*. Cambridge, MA: Harvard University Press.
Kimball Smith, Alice. 1965. *A peril and a hope: The scientists' movement in America 1945–47*. Chicago: University of Chicago Press.
Kinney, Aimee, and Thomas Leschine. 2002. A procedural evaluation of an analytic-deliberative process: The Columbia River comprehensive impact assessment. *Risk Analysis* 22:83–100.
Kitcher, Philip. 1993. *The advancement of science*. New York: Oxford University Press.
———. 1997. An argument about free inquiry. *Nous* 31:279–306.
———. 2001. *Science, truth, and democracy*. New York: Oxford University Press.
Kleinman, Daniel Lee. 1995. *Politics on the endless frontier: Postwar research policy in the United States*. Durham, NC: Duke University Press.
Kosso, Peter. 1989. Science and objectivity. *Journal of Philosophy* 86:245–57.
Krimsky, Sheldon. 1982. *Genetic alchemy: The social history of the recombinant DNA controversy*. Cambridge, MA: MIT Press.
———. 1991. *Biotechnics and society: The rise of industrial genetics*. New York: Praeger.
———. 2000. *Hormonal chaos: The scientific and social origins of the environmental endocrine hypothesis*. Baltimore: Johns Hopkins University Press.

———. 2003. *Science in the private interest: Has the lure of profits corrupted biomedical research?* Lanham, MD: Rowman and Littlefield.

Krimsky, Sheldon, and Roger Wrubel. 1998. *Agricultural biotechnology and the environment: Science, policy, and social issues.* Urbana: University of Illinois Press.

Kuhn, Thomas. 1962. *The structure of scientific revolutions.* Chicago: University of Chicago Press.

———. 1977. Objectivity, value judgment, and theory choice. In *The essential tension,* ed. Thomas Kuhn, 320–39. Chicago: University of Chicago Press.

Lacey, Hugh. 1999. *Is science value-free? Values and scientific understanding.* New York: Routledge.

———. 2005. *Values and objectivity in science: The current controversy about transgenic crops.* Lanham, MD: Rowman and Littlefield.

Laird, Frank. 1993. Participatory analysis, democracy, and technological decision-making. *Science, Technology, and Human Values* 18:341–61.

Landy, Marc K., Marc J. Roberts, and Stephen R. Thomas. 1990. *The Environmental Protection Agency: Asking the wrong questions.* New York: Oxford University Press.

Lapp, Ralph E. 1965. *The new priesthood: The scientific elite and the uses of power.* New York: Harper & Row.

Laudan, Larry. 1984. *Science and values: The aims of science and their role in scientific debate.* Berkeley and Los Angeles: University of California Press.

———. 2004. The epistemic, the cognitive, and the social. In *Science, values, and objectivity,* ed. Peter Machamer and Gereon Wolters, 14–23. Pittsburgh: University of Pittsburgh Press.

Leach, James. 1968. Explanation and value neutrality. *British Journal of Philosophy of Science* 19:93–108.

Leiserson, Avery. 1965. Scientists and the policy process. *American Political Science Review* 59:408–16.

Levi, Isaac. 1960. Must the scientist make value judgments? *Journal of Philosophy* 57:345–57.

———. 1962. On the seriousness of mistakes. *Philosophy of Science* 29:47–65.

Lloyd, Elisabeth. 1995. Objectivity and the double standard for feminist epistemologies. *Synthese* 104:351–81.

Longino, Helen. 1990. *Science as social knowledge.* Princeton, NJ: Princeton University Press.

———. 1996. Cognitive and non-cognitive values in science: Rethinking the dichotomy. In *Feminism, science, and the philosophy of science,* ed. Lynn Hankinson Nelson and Jack Nelson, 39–58. Dordrecht: Kluwer.

———. 2002. *The fate of knowledge.* Princeton, NJ: Princeton University Press.

Lowrance, William. 1976. *Of acceptable risk: Science and the determination of safety.* Los Altos, CA: William Kaufmann.

Lübbe, Hermann. 1986. Scientific practice and responsibility. In *Facts and values,* ed. M. C. Doeser and J. N. Kray, 81–95. Dordrecht: Martinus Nijhoff.

Lynn, Frances, and George Busenberg. 1995. Citizen advisory committees and environmental policy: What we know, what's left to discover. *Risk Analysis* 15:147–62.

MacCabe, Colin. 1998. Foreword. *Critical Quarterly* (London) 40:1–2.
Machamer, Peter, and Heather Douglas. 1998. How values are in science. *Critical Quarterly* (London) 40:29–43.
———. 1999. Cognitive and social values. *Science and Education* 8:45–54.
MacLean, Douglas, ed. 1986. *Values at risk.* Totowa, NJ: Rowan and Allanheld.
Maker, William. 1994. Scientific autonomy, scientific responsibility. In *Professional ethics and social responsibility*, ed. Daniel E. Wueste, 219–41. London: Rowman and Littlefield.
Malcolm, Norman. 1951. The rise of scientific philosophy. *The Philosophical Review* 60:582–86.
Margolis, Howard. 1996. *Dealing with risk: Why the public and the experts disagree on environmental issues.* Chicago: University of Chicago Press.
Mayo, Deborah, and Rachelle Hollander, eds. 1991. *Acceptable evidence: Science and values in risk management.* New York: Oxford University Press.
McCumber, John. 2001. *Time in the ditch: American philosophy and the McCarthy era.* Evanston, IL: Northwestern University Press.
McMullin, Ernan. 1983. Values in science. In *Proceedings of the 1982 biennial meeting of the Philosophy of Science Association*, vol. 1, ed. Peter D. Asquith and Thomas Nickles, 3–28. East Lansing, MI: Philosophy of Science Association.
———. 2000. Values in science. In *A companion to the philosophy of science*, ed. W. H. Newton-Smith, 550–60. Malden, MA: Blackwell.
Megill, Alan. 1994. Introduction: Four senses of objectivity. In *Rethinking Objectivity*, ed. Alan Megill, 1–20. Durham, NC: Duke University Press.
Merton, Robert K. 1938/1973. Science and the social order. Reprinted in *The sociology of science*, ed. Norman W. Storer, 254–66. Chicago: University of Chicago Press. Originally published in *Philosophy of Science* 5:321–37.
———. 1942/1973. The normative structure of science. Reprinted in *The sociology of science*, ed. Norman W. Storer, 267–78. Chicago: University of Chicago Press. Originally published in *Journal of Legal and Political Sociology* 1:115–26.
Mitchell, Sandra. 2004. The prescribed and proscribed values in science policy. In *Science, values, and objectivity*, ed. Peter Machamer and Geron Wolters, 245–55. Pittsburgh: University of Pittsburgh Press.
Mullins, Nicholas C. 1981. Power, social structure, and advice in American science: The United States national advisory system: 1950–1972. *Science, Technology, and Human Values* 7:4–19.
Mutz, Diana C. 2006. *Hearing the other side: Deliberative versus participatory democracy.* New York: Cambridge University Press.
Nagel, Ernest. 1961. *The structure of science: Problems in the logic of scientific explanation.* New York: Harcourt, Brace & World.
Nagel, Thomas. 1986. *The view from nowhere.* New York: Oxford University Press.
National Academy of Sciences. 1975. *Decision making for regulating chemicals in the environment.* Washington, DC: National Academy Press.
National Research Council. 1983. *Risk assessment in the federal government: Managing the process.* Washington, DC: National Academy Press.

———. 1994. *Science and judgment in risk assessment.* Washington, DC: National Academy Press.

———. 1996. *Understanding risk: Informing decisions in a democratic society.* Washington, DC: National Academy Press.

Nelkin, Dorothy. 1975. The political impact of technical expertise. *Social Studies of Science* 5:35–54.

Neumann, D. A., and S. S. Olin. 1995. Urinary bladder carcinogenesis: A working group approach to risk assessment. *Food and Chemical Toxicology* 33:701–4.

Nozick, Robert. 1998. Invariance and objectivity. *Proceedings and Addresses of the APA* 72:21–48.

Nye, Mary Jo. 1972. *Molecular reality.* New York: American Elsevier.

Office of Management and Budget. 2002. Guidelines for ensuring and maximizing the quality, objectivity, utility, and integrity of information disseminated by federal agencies; republication. *Federal Register* 67:8452–60.

———. 2005. Final information quality bulletin for peer review. *Federal Register* 70:2664–77.

Parliament of Science. 1958. 1958 Parliament of Science. *Science* 127:852–58.

Parsons, Keith. 2003. From Gish to Fish: Personal reflections on the science wars. In *The science wars: Debating scientific knowledge and technology,* ed. Keith Parsons, 9–16. Amherst, NY: Prometheus Books.

Paul, Ellen Frankel, Fred Miller, and Jeffrey Paul, eds. 1999. *Responsibility.* Cambridge: Cambridge University Press.

Perl, Martin. 1971. The scientific advisory system: Some observations. *Science* 173:1211–15.

Perrin, Jean. 1913. *Atoms.* Trans. D. L. Hammick, 1916. Woodbridge, CT: Ox Bow Press.

Porter, Theodore. 1992. Quantification and the accounting ideal in science. *Social Studies of Science* 22:633–52.

———. 1995. *Trust in numbers: The pursuit of objectivity in science and public life.* Princeton, NJ: Princeton University Press.

Presidential/Congressional Commission on Risk Assessment and Risk Management. 1997a. *Framework for Environmental Health Risk Management,* vol. 1.

———. 1997b. *Risk assessment and risk management in regulatory decision-making,* vol. 2.

Price, Don K. 1954/1962. *Government and science.* Repr., New York: Oxford University Press.

———. 1965. *The scientific estate.* Cambridge, MA: Harvard University Press.

———. 1969. Purists and politicians. *Science* 163:25–31.

Proctor, Robert. 1991. *Value-free science? Purity and power in modern knowledge.* Cambridge, MA: Harvard University Press.

Pursell, Carroll W., Jr. 1965. The anatomy of failure: The science advisory board, 1933–1935. *Proceedings of the American Philosophical Society* 109:342–51.

———. 1979. Science agencies in World War II: The OSRD and its challengers. In *The sciences in the American context: New perspectives,* ed. Nathan Reingold, 359–78. Washington, DC: Smithsonian Institution Press.

Quine, W. V. O. 1992. *The pursuit of truth*. Cambridge, MA: Harvard University Press.

Rawls, John. 1971. *A theory of justice*. Cambridge, MA: Harvard University Press.

Reichenbach, Hans. 1938. *Experience and prediction*. Chicago: University of Chicago Press.

———. 1951. *The rise of scientific philosophy*. Berkeley and Los Angeles: University of California Press.

Reingold, Nathan, ed. 1979. *The sciences in the American context: New perspectives*. Washington, DC: Smithsonian Institution Press.

Reisch, George. 2005. *How the cold war transformed philosophy of science: To the icy slopes of logic*. New York: Cambridge University Press.

Renn, Ortwin. 1999. A model for an analytic-deliberative process in risk management. *Environmental Science and Technology* 33:3049–55.

Resnick, David B. 1998. *The ethics of science*. New York: Routledge.

Rhodes, Richard. 1986. *The making of the atomic bomb*. New York: Simon and Schuster.

Richardson, Alan. 2003. Logical empiricism, American pragmatism, and the fate of scientific philosophy in North America. In *Logical empiricism in North America*, ed. Alan Richardson and Gary Hardcastle, 1–24. Minneapolis: University of Minnesota Press.

Richardson, Alan, and Gary Hardcastle, eds. 2003. *Logical empiricism in North America*. Minneapolis: University of Minnesota Press.

Rooney, Phyllis. 1992. On values in science: Is the epistemic/non-epistemic distinction useful? In *Proceedings of the 1992 biennial meeting of the Philosophy of Science Association*, vol. 2, ed. David Hull, Micky Forbes, and Kathleen Okruhlik, 13–22. East Lansing, MI: Philosophy of Science Association.

Rosenstock, Linda, and Lore Jackson Lee. 2002. Attacks on science: The risks to evidence-based policy. *American Journal of Public Health* 92:14–18.

Rowe, Gene, and Lynn Frewer. 2000. Public participation methods: A framework for evaluation. *Science, Technology, and Human Values* 25:3–29.

Ruckelhaus, William. 1983. Risk, science, and public policy. *Science* 221:1026–28.

———. 1985. Risk, science, and democracy. *Issues in Science and Technology* 1:19–38.

Rudner, Richard. 1953. The scientist *qua* scientist makes value judgments. *Philosophy of Science* 20:1–6.

Santer, B. D., T. M. L. Wigley, G. A. Meehl, M. F. Wehner, C. Mears, M. Schabel, F. J. Wentz, et al. 2003. Influence of satellite data uncertainties on the detection of externally forced climate change. *Science* 300:1280–84.

Sapolsky, Harvey M. 1968. Science advice for state and local government. *Science* 160:280–84.

———. 1979. Academic science and the military: The years since the Second World War. In *The sciences in the American context: New perspectives*, ed. Nathan Reingold, 379–99. Washington, DC: Smithsonian Institution Press.

Sarewitz, Daniel. 1996. *Frontiers of illusion: Science, technology, and the politics of progress*. Philadelphia: Temple University Press.

Schrecker, Ellen. 1986. *No ivory tower: McCarthyism and the universities.* New York: Oxford University Press.
Sclove, R. E. 2000. Town meetings on technology: Consensus conferences as democratic participation. In *Science, technology, and democracy,* ed. Daniel Lee Kleinman, 33–48. Albany, NY: SUNY Press.
Scriven, Michael. 1974. The exact role of value judgments in science. In *PSA 1972: Proceedings of the 1972 biennial meeting of the Philosophy of Science Association,* ed. Kenneth F. Schaffner and Robert S. Cohen, 219–47. Dordrecht: Reidel.
Segerstråle, Ullica. 2000. Science and science studies: Enemies or allies? In *Beyond the science wars: The missing discourse about science and society,* ed. Ullica Segerstråle, 1–40. Albany, NY: SUNY Press.
Shrader-Frechette, Kristin. 1991. *Risk and rationality.* Berkeley and Los Angeles: University of California Press.
———. 1993. *Burying uncertainty: Risk and the case against geological disposal of nuclear waste.* Berkeley and Los Angeles: University of California Press.
———. 1994. *Ethics of scientific research.* Lanham, MD. Rowman and Littlefield.
Silbergeld, Ellen. 1991. Risk assessment and risk management: An uneasy divorce. In *Acceptable evidence: Science and values in risk management,* ed. Deborah Mayo and Rachelle Hollander, 99–114. New York: Oxford University Press.
Sinsheimer, Robert. 1979. The presumptions of science. In *Limits of scientific inquiry,* ed. Gerald Holton and Robert Morison, 23–35. New York: W. W. Norton.
Slovic, Paul. 1987. Perception of risk. *Science* 236:280–85.
Smith, Bruce L. R. 1990. *American science policy since World War II.* Washington, DC: The Brookings Institution.
———. 1992. *The advisors: Scientists in the policy process.* Washington, DC: The Brookings Institution.
Sokal, Alan. 1998. What the Sokal Affair does and does not prove. In *A house built on sand: Exposing postmodernist myths about science,* ed. Noretta Koertge, 9–22. New York: Oxford University Press.
Solomon, Miriam. 2001. *Social empiricism.* Cambridge, MA: MIT Press.
Thomson, Judith Jarvis. 1971. A defense of abortion. *Philosophy and Public Affairs* 1:47–66.
Uebel, Thomas. 2003. Philipp Frank's history of the Vienna Circle: A programmatic retrospective. In *Logical empiricism in North America,* ed. Alan Richardson and Gary Hardcastle, 149–69. Minneapolis: University of Minnesota Press.
Union of Concerned Scientists. 2004. Scientific integrity in policymaking: An investigation into the Bush administration's misuse of science. http://www.uscsusa.org.
U.S. Congress. House. Committee on Government Reform—Minority Staff. 2003. Politics and science in the Bush administration. http://www.reform.house.gov/min.
Viscusi, Kip. 1992. *Fatal tradeoffs: Public and private responsibilities for risk.* New York: Oxford University Press.
Walker, Eric A. 1967. National science board: Its place in national policy. *Science* 156:474–77.

Weinberg, Alvin. 1972. Science and trans-science. *Minerva* 10:209–22.
Westwick, Peter J. 2003. *The national labs: Science in an American system, 1947–1974.* Cambridge, MA: Harvard University Press.
Withers, R. F. J. 1952. The rise of scientific philosophy. *British Journal for the Philosophy of Science* 2:334–37.
Wolfle, Dael. 1959. *Science and public policy.* Lincoln: University of Nebraska Press.
Wolpert, Lewis. 1992. *The unnatural nature of science.* Cambridge, MA: Harvard University Press.
Wood, Robert. 1964. Scientists and politics: The rise of an apolitical elite. In *Scientists and national policy-making,* ed. Robert Gilpin and Christopher Wright, 41–72. New York: Columbia University Press.
Wylie, Alison. 2003. Why standpoint matters. In *Science and other cultures: Issues in philosophies of science and technology,* ed. Robert Figueroa and Sandra Harding, 26–48. New York: Routledge.
Zachary, G. Pascal. 1997. *Endless frontier: Vannevar Bush, engineer of the American century.* New York: Free Press.

INDEX

advisory boards, 42
Advisory Committee on Reactor Safeguards, 34
aesthetic values, 92
American Association for the Advancement of Science (AAAS), 36, 52–53
analogical reasoning, 170–71
analytic-deliberative processes, 159–67
antiballistic missile (ABM) program, 39
Aristotle, 107
Asilomar conference, 77
atomic bomb, 31–32, 75, 77–78, 83–84, 186n9
Atomic Energy Act, 34
Atomic Energy Commission (AEC), 34–36
authority: political, 134; of science, 4–8, 82, 106, 135
autonomy of science: hypothesis evaluation and, 53–56; moral exemption and, 76; problems with concept of, 7–8, 13; and soundness of science, 10–11; value-free ideal and, 7, 16–17, 45–46, 60–63
Avogadro's number, 120
Ayer, A. J., 47, 49

Bacon, Francis, 46
Bazelon, David, 138
Bethe, Hans, 38, 77
blame, 68
Bohr, Niels, 90
Breyer, Stephen, 12
Bridgman, Percy, 75–77
British Journal for the Philosophy of Science, 49
Bureau of the Census, 25
Burying Uncertainty (Shrader-Frechette), 17
Busenberg, George, 164
Bush, George W., 12
Bush, Vannevar, 29–32, 34

canons of inference, 55, 58, 90
carcinogenicity, 143–46
Carnap, Rudolph, 47

causation, and moral responsibility, 67–68
certainty, 2–3, 8–9. *See also* uncertainty
Chadwick, James, 83–84
Churchman, C. West, 50–52, 55, 57, 65, 66, 183n12
Clean Air Act, 42, 138
Clean Water Act, 42
climate change, 130–31, 150, 166
coercion, 67
cognitive values: case study illustrating, 108–12; defined, 93–94; direct role of, in research, 98–103, 110–12; indirect role of, in research, 106–8, 111–12; social values reinforcing, 110; theory acceptance and, 107, 188n20. *See also* epistemic values
cold war, 48–49
collaborative analysis. *See* analytic-deliberative processes
competence, 67
Compton, Karl, 27–28, 30, 182n19
Conant, James, 30
concordant objectivity, 126–27, 130–31
confirmation studies, 48, 50
Congress. *See* U.S. Congress
consensus conferences, 168–69
consequences of error: case study illustrating, 108–12; foresight and, 83–84; moral responsibility and, 54, 58–59, 66–72, 80–83; multiple, 54; social/ethical values and, 58–59, 92; uncertainty and, 70, 78, 81; unintended, 68–70, 77–78; values and, 103–8, 162
consequentialist calculus, 170
consistency, 93, 94
Consumer Product Safety Act, 138
Consumer Product Safety Commission, 42
contract research grants, 29–31, 33
convergent objectivity, 119–21, 130–31
Cranor, Carl, 19–20
creationism, 21, 179n5, 180n13

205

Daniels, Josephus, 25
Daston, Lorraine, 115
Data Quality Act, 12
Decision Making for Regulating Chemicals in the Environment (National Research Council), 137
Defense Science Board, 35
deliberation: about values, 167–72; analytic-deliberative processes in public policy, 159–67
democracy: deliberative vs. participatory, 191n1; public policymaking role in, 158; scientists' role in, 66, 134–36
Democratic Advisory Council, 37
Democratic National Committee, 37
Department of Agriculture, 25
Department of Defense (DOD), 35, 39
Department of Energy (DOE), 166
Department of State, 37
DES. *See* diethylstilbestrol
detached objectivity, 122–24, 130–31, 149
Dewey, John, 47, 191n2
diethylstilbestrol (DES), 108–12, 119
dioxin, 152
disinterestedness, 46–47
DNA, 77, 84, 107, 118
DuBridge, Lee, 34, 39
Ducasse, C. J., 183n12
Duncan, Francis, 34
Dupree, Hunter, 25, 28

Edison, Thomas, 25
Einstein, Albert, 90, 186n9
Eisenhower, Dwight D., 35–36
Environmental Protection Agency (EPA), 42, 138, 145–47, 149, 152, 154
epistemic values: cognitive vs., 93–94; instances of, 55, 64, 94; as necessary conditions for science, 187n8; nonepistemic vs., 89–91; as only legitimate scientific values, 55–56, 58, 60–62, 64. *See also* cognitive values
errors, types of, 104–6. *See also* consequences of error
ether, 2
ethical review boards. *See* internal review boards
ethics: knowledge and, 48; philosophical, 169–70; philosophy of science and, 182n1; research, 9, 71–72, 74, 76, 92, 99–100, 182n1, 187n11. *See also* moral responsibility; social and ethical values; values in science
Ethics of Scientific Research (Shrader-Frechette), 17–18
ethos of science, 46–47
evidence: risk assessment and, 146; sound science–junk science dispute and, 5, 150; standards of, 10, 12; sufficiency of, 50–51, 97, 148, 162; values and, 57–58, 87, 96–97, 102, 106, 122, 153, 173
experimental runs, acceptance/rejection of, 105
experts: balance of, on advisory committees, 41; credibility of, 5; dueling, 42, 140, 149–50, 164, 167; and evidence, 10; incompetent, 151; outside areas of expertise, 151
explanatory power, 93, 106–7

FACA. *See* Federal Advisory Committee Act
fact-value distinction, 57, 64
false negatives, 104–6, 187n18
false positives, 104–6, 187n18
Federal Advisory Committee Act (FACA), 39, 41, 43, 135, 137
Federal Council for Science and Technology (FCST), 36–37
Federal Rules of Evidence, 10
Federal Water Pollution Control Act, 138
Feinberg, Joel, 68–69
feminist scholarship, 6, 18, 91, 101
Fermi, Enrico, 34
Feyerabend, Paul, 184n26
Fine, Arthur, 126
Fiorino, Daniel, 158
fission, 83–84, 186n7, 186n9
Food and Drug Administration (FDA), 108–9
forbidden knowledge, 72, 187n12
Frank, Philip, 47, 50, 52–53
fruitfulness, 93, 107
Frye rule, 10, 12
Fuller, Steve, 179n5
funding, 4, 21, 99

Gaa, James, 62, 64
Galileo Galilei, 79, 113
Galileo's Revenge (Huber), 10
Garwin, Richard, 40
General Advisory Committee (GAC), 34
Geological Survey, 25
global warming. *See* climate change
Golden, William, 35
government. *See* public policy; science advice
Great Depression, 27
Greenberg, Daniel, 32
Gross, Paul, 6, 179n1
Guidelines for Carcinogen Risk Assessment (Environmental Protection Agency), 145–47, 154

Hacking, Ian, 118
Hahn, Otto, 83
Hale, George Ellery, 26
Hanford, Washington, groundwater contamination in, 166
Hardimon, Michael, 186n4
Hart, Roger, 6
Heil, John, 16, 97
Hempel, Carl, 48, 58–61, 183n14, 184n22
Herrick, Charles, 11
Hewlett, Richard, 34
Higher Superstition (Gross and Levitt), 6
Hippel, Frank von, 41
Hitchens, Christopher, 179n2

hoaxing, 7
Hoffmann, Roald, 64
honest science, 19
Hoover, Herbert, 26
Hormonal Chaos (Krimsky), 19
hormones, 108–12
Huber, Peter, 10–11, 64, 135
human-world interactions, objectivity in, 118–21
hypotheses: certainty and, 2; probabilities assigned for, 53–54, 183n15; scientific community as context for evaluation of, 53–56; values influential in evaluation of, 51–54, 58–59. *See also* inference
hypothesis testing, 2

impartiality of science, 17
inductive gaps, 2, 96
inductive risk, 58–59
inference: canons of, 55, 58, 90; guidelines on, 142–48; in risk assessment, 142–48; values and, 51. *See also* hypotheses
Institute for the Unity of Science, 52
integrity of science: clarity and explicitness as aid to, 152–54, 164; disagreements among scientists and, 152; risk assessment and, 147; sound vs. junk science, 4–5, 8–13, 148–54; value-free ideal and, 134; values' role limited to assure, 96, 98, 100–102, 123, 136, 148–49, 153–55, 176
intelligent design, 21, 179n5, 180n13
intention, 68
interactive objectivity, 127–28, 131, 154, 160
Intergovernmental Panel on Climate Change (IPCC), 131, 166
internal consistency, 94
internal review boards, 74, 84
interpretation of data, 106
intersubjectivity, 126–28
invariance, 189n5
Is Science Value-Free? (Lacey), 16, 63
ivory-billed woodpeckers, 129–30

Jamieson, Dale, 11
Jeffrey, Richard, 53–54, 59, 63, 85, 183n14, 183n15
Jewett, Frank, 30, 32
Johnson, Lyndon, 38–39
Joint Research and Development Board, 34–35
junk science, 4–5, 10–13, 148–54
just-so stories, 107

Kelly, Thomas, 16
Kennedy, John F., 36–37
Kepler, Johannes, 107
Kevles, Daniel, 32
Killian, James, 36
Kipling, Rudyard, 107
Kissinger, Henry, 39–40
Kistiakowsky, George, 77

knowledge: ethics and, 48; forbidden, 72, 187n12; as goal of science, 94; realism and, 189n2, 189n3; social construction of, 5–6; social value of, 76–77
Korean War, 35
Krimsky, Sheldon, 19
Kuhn, Thomas, 10–11, 60–62, 64, 65, 184n26

Lacey, Hugh, 16–17, 63, 91
Laudan, Larry, 61–62, 93–94, 184n25
Leach, James, 62, 184n27
Lee, Lore Jackson, 135–36
Levi, Isaac, 55–56, 58, 62, 63, 66, 90, 183n17, 186n2
Levitt, Norman, 6, 179n1
Lilienthal, David E., 34
Lloyd, Lisa, 122
logical empiricists, 47
Longino, Helen, 18–20, 57, 90–91, 110, 128, 172, 186n2
Lowrance, William, 140–41
Lübbe, Hermann, 75
Lusitania (ship), 25
Lysenko affair, 79, 102, 113

MacLean, Douglas, 20
Malcolm, Norman, 49
Manhattan Project, 31–32, 78
manipulable objectivity, 118–19, 121, 131, 154
Marxism, 49
MAUD report, 31
McMullin, Ernan, 63, 66, 90
Meitner, Lise, 83
Merton, Robert, 46–47
methodology, 99–102, 104–5
Millikan, Robert, 105
"Moral Autonomy and the Rationality of Science" (Gaa), 62
moral responsibility: authority and, 82; concept of, 67–71; and consequences of error, 54, 58–59, 66–72, 80–83; exemption from, 75–79; limits to, 82–85; science advising and, 80–82; of scientists, 67, 71–86, 175–76; uncertainty and, 81, 85. *See also* ethics; social and ethical values; values in science

Nagel, Ernst, 56–58
National Academy of Sciences (NAS), 23, 25–28, 30
National Advisory Committee on Aeronautics (NACA), 26–27, 29–30, 32, 34
National Aeronautics and Space Administration (NASA), 36
National Bureau of Standards, 25
National Cancer Institute, 29
National Defense Research Committee (NDRC), 30–31
National Environmental Protection Act, 42

208 • INDEX

National Institutes of Health (NIH), 34, 35
national laboratories, 32
National Research Council (NRC), 26–27, 42, 137, 140, 142–44, 147, 149, 155, 157, 159–62, 190n5
National Research Fund, 26
National Resources Board (NRB), 27–28, 181n6
National Resources Committee, 181n6
National Science Foundation (NSF), 33–37, 40, 52
Naval Consulting Board (NCB), 25–26
Naval Observatory, 32
Naval Research Laboratory, 26
negligence, 68–70
neo-Thomists, 182n5
Neurath, Otto, 47
neutrality, 17, 123–24
New Deal, 27
Nixon, Richard, 39–40, 137
"The Normative Structure of Science" (Merton), 46
Nozick, Robert, 189n5
NRC. *See* National Research Council
nuclear physics, 83–84
nuclear weapons. *See* atomic bomb
Nye, Mary Jo, 120

objectivity, 115–32; complexity of concept of, 115–16, 132; concordant, 126–27, 130–31; convergent, 119–21, 130–31; degrees of, 117, 128–29; detached, 122–24, 130–31, 149; endorsement as characteristic of, 116–17; in human-world interactions, 118–21; interactive, 127–28, 131, 154, 160; invariance and, 189n5; manipulable, 118–19, 121, 131, 154; procedural, 125–26, 131, 160; processes for, 117–18; realism and, 188n2, 189n3; in social processes, 125–29, 190n10; social/ethical values and, 115; strong sense of, 117, 119; synergism among bases for, 129–32; in thought processes, 121–24; traditional notion of, 188n2; trust as characteristic of, 116–17; truth and, 117; value-free, 122–23, 132, 149; value-neutral, 123–24, 131, 154; values and, 15, 51–52, 122–26
Occupational Safety and Health Act, 138
Occupational Safety and Health Administration, 42
Office of Defense Mobilization (ODM), 35
Office of Management and Budget (OMB), 12, 179n6
Office of Naval Research (ONR), 34, 181n10
Office of Science Advisor, 37
Office of Science and Technology (OST), 36–38
Office of Scientific Research and Development (OSRD), 31–33
Oppenheim, Paul, 48
Oppenheimer, Robert, 34

paradigms, 60
Parliament of Science, 36

participatory research. *See* analytic-deliberative processes
Perl, Martin, 40
Perrin, Jean, 120
philosophy, and values deliberation, 169–70
philosophy of science: emergence of, 183n18; focus of, 44, 48, 53, 56; Kuhn's impact on, 61; and social character of science, 128; and societal role of science, 44; theory-practice distinction in, 16; and value-free ideal, 14, 16–18, 44, 47–50, 53–56, 60–64; and values in science, 16–21, 56–60
Philosophy of Science Association, 52
Philosophy of Science (journal), 48, 51, 183n12, 185n28
plutonium bomb, 77–78
policy. *See* public policy
politicized science, 12–13, 113
Porter, Theodore, 125–26
praise, 68
predictive competency, 94
predictive precision, 93, 107
Presidential Science Advisory Committee (PSAC), 35–40, 42, 137
Presidential/Congressional Commission on Risk Assessment and Risk Management, 157
Primack, Joel, 41
probability, 53–54, 183n15
problem of forbidden knowledge, 72, 187n12
procedural objectivity, 125–26, 131, 160
Proctor, Robert, 46
pseudoscience, 10
public: accountability to, 16; benefits for science advisors from participation of, 173–74; diversity of values in, 172–73; policymaking role of, 16, 157–74; and values deliberations, 167–72
Public Health Service, 25
public policy: analytic-deliberative processes in, 159–67; complexity and, 9; public role in, 16, 157–74; science's role in, 3–4, 8–13, 15–16, 19–20, 45, 62–63, 133–48, 155; sound vs. junk science and, 4–5, 8–13, 148–54; value-free ideal and, 135. *See also* science advice
pure science, 113

quantification, 125–26
Quine, W. V. O., 127

Rabi, I. I., 34
radioactivity, 2
realism, 188n2, 189n3
recklessness, 68–70
recombinant DNA, 77, 84
Regional Citizen's Advisory Council (RCAC), 164
Regulating Toxic Substances (Cranor), 19–20
regulation: growth of, 138; risk analysis and, 140; uncertainty and, 139

regulatory agencies: decisionmaking authority in, 134, 158; growth of, 138; inference guidelines of, 142–48; public policymaking role and, 158; and science advice, 134–35
Reichenbach, Hans, 48–49
research: contract research grants, 29–31, 33; decision points and values in, 88–89, 95, 98–108; ethics of, 9, 71–72, 74, 76, 92, 99–100, 182n1, 187n11
Resnick, David, 182n1
Resource Conservation and Recovery Act, 138
The Rise of Scientific Philosophy (Reichenbach), 48–49
risk analysis: components of, 139–41; decision points and values in, 161–62; emergence of, 138–39; and policymaking role of science, 139–48; regulation and, 140; values and, 140–48, 154–55
Risk and Rationality (Shrader-Frechette), 17
risk assessment, 17, 21; analytic-deliberative processes in, 159–67; case-specific considerations in, 146–49, 163; defined, 139–41; inference in, 142–47; public policymaking role in, 158–64; risk management vs., 140–41, 143–44, 190n5, 190n6; steps in, 140; uncertainty and, 142, 161; values and, 136–37, 141–42, 147–48, 154–55, 162–63
risk characterization, 159–61
risk management, 21; defined, 139–41; risk assessment vs., 140–41, 143–44, 190n5, 190n6; values and, 140–41
role responsibilities, 72–75; aspects of, 72; defined, 72; general vs., 73–75, 186n4; in science, 73–75
Rooney, Phyllis, 90, 110
Roosevelt, Franklin D., 27, 29–30, 32
Rosenstock, Linda, 135–36
Ruckelhaus, William, 190n6
Rudner, Richard, 50–55, 57, 59, 63, 65, 66, 80, 85, 106, 183n17
Rutherford, Ernest, 83

Safe Drinking Water Act, 138
Sapolsky, Harvey M., 181n10
Schaffner, Ken, 185n28
science: accepted vs. state-of-the-art, 12; authority of, 4–8, 82, 106, 135; complexity of, 9; confinement of, to scientific community, 53–56, 60–61, 82–83; criticisms of, 4–8; diversity in community of, 172–73; ethos of, 46–47; funding of, 4, 21; generality of, 3; government and (*see* public policy; science advice); honest, 19; impartiality of, 17; neutrality of, 17; politicized, 12–13, 113; presumptions in, 101; public funding of, 4, 99; public policy role of, 3–4, 8–13, 15–16, 19–20, 45, 62–63, 133–48, 155; reliability of, 1–3, 8, 13, 94–96; responsibility sharing by, 74, 84–85, 157, 171–72; self-corrective social mechanisms in, 57; and sexism, 6; social character of, 128; societal role of, 4–5, 7–8, 13–14, 44–45, 47–48, 50, 172; sound vs. junk, 4–5, 8–13, 148–54; value of, 95, 102. *See also* autonomy of science; integrity of science; research; science advice; Science Wars; scientists; value-free ideal; values in science
science advice, 14, 23–43; Civil War and, 23; controversy and regulation pertaining to (1965–1975), 38–43; effectiveness of, 135; executive branch and, 36, 38–41; importance of, 43; increasing demand for (1970s), 138; moral responsibility concerning, 80–82; post–World War II to 1970, 33–38; pre–World War II, 24–29; public nature of, 40–41, 137–38, 163; public participation in, 173–74; role of, 37, 43, 134–35; scientists' participation in, 37–38; tensions in, 28, 38–39; World War II and, 14, 29–33
science advisory boards, 42
Science Advisory Board (SAB), 27–29
Science Advisory Committee (SAC), 35
Science and Engineering Ethics (journal), 182n1
"Science and Human Values" (Hempel), 58
Science and Values (Laudan), 61
Science as Social Knowledge (Longino), 18
science education, 21, 179n5
Science (magazine), 12
Science: The Endless Frontier (Bush), 32
Science Wars, 4–8, 13, 45, 179n5
Scientific Monthly (magazine), 52
scientists: diversity among, 172–73; foresight and responsibility of, 83–85; moral exemption of, 75–79; moral responsibilities of, 66–86; political views of, 41; role responsibilities of, 72–75; and unintended consequences, 71–72, 77–78
scope, 93, 107
Scriven, Michael, 62
Seaborg, Glenn, 34
Segerstråle, Ullica, 7, 8
sexism, 6
Shrader-Frechette, Kristen, 17–18, 20, 182n1
significance, statistical, 104, 187n18
simplicity, 93, 107
Smith, Bruce L. R., 39
social and ethical values: case study illustrating, 108–12; cognitive values reinforcing, 110; and consequences of error, 58–59, 92; defined, 92–93; direct role of, in research, 98–103, 110–11; indirect role of, in research, 103–6, 111–12, 148–49; integrity of science and, 148–49, 154; objectivity and, 115; risk assessment and, 136–37, 147–48, 154–55, 162–63; risk management and, 140–41; in science, 89–91; scientific enterprise among, 95–96; scientific judgments and, 136. *See also* ethics; moral responsibility

social constructivism, 5–7
social processes, objectivity in, 125–29, 190n10
social sciences, 21
Social Text (journal), 7, 179n2
Sokal, Alan, 7
Solomon, Miriam, 172
sound science–junk science dispute, 4–5, 8–13, 148–54
Sputnik, 35–36
statistical significance, 104, 187n18
Strassmann, Fritz, 83
The Structure of Science (Nagel), 56
The Structure of Scientific Revolutions (Kuhn), 60
supersonic transport (SST), 40
Szilard, Leo, 186n9

theology, 90
theory vs. practice, 16
Thomson, Judith Jarvis, 170–71
thought, objectivity in, 121–24
Toxic Substances Control Act, 138
transformative criticism, 128
transparency, about values, 136, 153–56, 176–77
trans-science, 139
Trinity test, 77–78
Truman, Harry S., 33, 35
trust, 116–17
truth: epistemic values and, 93–94; objectivity and, 117
"Turns in the Evolution of the Problem of Induction" (Hempel), 59
Tuskegee syphilis experiments, 187n11

uncertainty: case study illustrating, 108–12; consequences of error and, 70, 78, 81; data evaluation and, 149–50; generality of science and, 3; moral responsibility and, 81, 85; public policy and, 8–9; regulation and, 139; in risk analysis, 142, 161; values and, 105–6. *See also* certainty
Understanding Risk (National Research Council), 157, 159
unintended consequences: moral responsibility for, 68–70; scientists and, 71–72, 77–78
universal reflexivity, 170
The Unnatural Nature of Science (Wolpert), 6, 64
U.S. Congress, 37–38, 40–42, 134

Valdez, Alaska, marine oil trade in, 164–65
"Valuation and Acceptance of Scientific Hypotheses" (Jeffrey), 53
value-free ideal, 44–65; acceptable vs. unacceptable values and, 89–91; ambivalence over, 56–60; autonomy and, 7, 16–17, 45–46, 60–63; critique of, 15, 20, 103, 175; defined, 45, 180n8; development of, 14; impartiality and, 16–17; and integrity of science, 134; moral exemption and, 79; neutrality and, 16–17; objectivity and, 122–23, 132, 149; origins of, 45–46; philosophy of science and, 14, 16–18, 44, 47–50, 53–56, 60–64; proponents of, 16–17, 53–56, 60–64; public policy and, 135
value-free objectivity, 122–23, 149
value-neutral objectivity, 123–24, 131, 154
values in science, 15, 87–114; acceptable vs. unacceptable, 89–92, 95, 186n2; aesthetic, 92; case study illustrating, 108–12; clarity and explicitness about, 136, 153–56, 176–77; cold war attitudes toward, 48–50; and consequences of error, 103–8, 162; direct role for, 87–88, 96–103, 110–13, 122, 148–49, 162–63; diversity of, 172–73; evidence and, 57–58, 87, 96–97, 102, 106, 122, 153, 173; facts and, 57, 64; funding and, 99; hiding of, 123; hypothesis evaluation and, 51–54, 58–59, 96–97; indirect role for, 87–88, 96–98, 103–8, 111–13, 148–49, 163; inference and, 51; instrumental vs. categorical values, 184n22; integrity of science and, 136; interpretation and, 106; limits on, 96, 98, 100–102, 123, 155, 176; methodology and, 99–102, 104–5; misuse of, 100–102, 110–13, 122; necessity of, 112–14; objectivity and, 51–52, 122–26; philosophy of science and, 16–21, 56–60; presumptions and, 101; pre–World War II notion of, 47; project selection and, 98–99; proponents of, 50–53, 56–60; in research process, 88–89, 95, 98–108; risk analysis and, 140–48, 154–55, 162–63; topology of, 89–95; uncertainty and, 105–6. *See also* cognitive values; epistemic values; ethics; moral responsibility; social and ethical values; value-free ideal
values, public deliberation on, 167–72
Van Fraassen, Bas, 186n2
Vietnam War, 38–39

Wallace, Henry A., 27
Weber, Max, 46
Weinberg, Alvin, 139
Wells, H. G., 186n7
"Why Buy That Theory?" (Hoffmann), 64
Wiesner, Jerome, 37
Withers, R. F. J., 49
Wolpert, Lewis, 6, 64
World War I, 25–26
World War II, 14, 29–33, 47